GENETICS - RESEARCH AND ISSUES

ALLELIC FORMS OF THE FMR1 GENE

FRAGILE X SYNDROME, PRIMARY OVARIAN INSUFFICIENCY AND TREMOR ATAXIA SYNDROME AMONG OTHERS

GENETICS - RESEARCH AND ISSUES

Additional books in this series can be found on Nova's website under the Series tab.

Additional e-books in this series can be found on Nova's website under the e-book tab.

GENETICS - RESEARCH AND ISSUES

ALLELIC FORMS OF THE FMR1 GENE

FRAGILE X SYNDROME, PRIMARY OVARIAN INSUFFICIENCY AND TREMOR ATAXIA SYNDROME AMONG OTHERS

MONTSERRAT MILÀ
EDITOR

nova publishers
New York

Library of Congress Cataloging-in-Publication Data

ISBN: 978-1-63321-914-4

Library of Congress Control Number: 2015931176

Published by Nova Science Publishers, Inc. † New York

Contents

Preface

This book provides an update of the different allelic forms of the *FMR1* gene.

The *FMR1* gene is an example of how a single gene can have different phenotypic effects, and since its discovery in 1991 it has been revealing new facets: Fragile X syndrome (FXS), Primary Ovarian Insufficiency (FXPOI), and Fragile X Tremor Ataxia Syndrome FXTAS, among others and from which we are continuously learning about new aspects of this gene.

Despite the lengthy journey made up to now, there still remains much to learn and we hope that more in depth knowledge of this gene will lead to new research into possible treatments for the different disorders related to the *FMR1*gene.

Albeit rare, fragile X syndrome is the most common cause of familial intellectual disability and the most commonly known single gene causing autism spectrum disorders. The incidence of this syndrome varies according to the populations; with 1/4,000 males and 1/6,000 females being affected, and 1/250 females and 1/800 males are carriers. The main clinical manifestations are intellectual disability, dimorphic traits and behavior disturbances. The syndrome is inherited as a dominant X-linked trait with reduced penetrance (80% for males and 30% for women). This syndrome is due to a functional loss of the *FMR1* gene product, Fragile X Mental retardation Protein (FMRP), and in most cases it is caused by a CGG repeats expansion in the *FMR1* promoter. The absence of the FMR1 protein, FMRP, is the cause of the intellectual disability in these patients. At present, there is no treatment, although several studies are ongoing in both human and animal models.

FXS patients present enormous functional impairment with a behavioral phenotype of hyperactivity and impaired attention, marked anxiety, with poor eye contact, affective liability, aggression, self-injurious behavior and autistic features. Since its first description the basic formulation of the fragile X behavioral phenotype has remained intact up to the present, with substantial confirmation of these basic findings in subsequent studies. Autism is one of the most recognized and severe behavioral abnormalities observed in males with FXS) and this syndrome is the leading known monogenic cause of autism. In individuals with the full mutation, approximately 30–50% meet full Diagnostic and Statistical Manual of Mental Disorders DSM-IV-TR criteria for autism with 60–74% fulfilling criteria for an autism spectrum disorder (ASD). Attention deficit and hyperactivity disorder (ADHD) is the most common diagnosable condition in FXS patients. In children with FXS, ADHD is characterized by more inattentiveness, restlessness, fidgetiness and impulsivity.

For many years it was thought that permutation (PM) carriers showed no clinical manifestations, only the risk of transmitting the disease to their offspring. In 1999 the presence of a PM became associated with FXPOI, and in 2000 a risk of late onset neurodegenerative disease Fragile X Tremor Ataxia Syndrome FXTAS was associated with male and female PM carriers. Lastly, in the years 2007-2009 a range of phenotypes associated with the PM were described. It is now accepted that FXPOI (MIM#311360) and FXTAS (MIM#300623) are different from FXS and are only associated with the *FMR1* gene PM.

Fragile X premature ovarian insufficiency is a clinical entity in which carrier PM females present early ovarian dysfunction, with menopause occurring 5 years earlier than non-carrier family members. It has been estimated that 2-6% of all women with premature ovarian failure and a normal karyotype have a PM. Therefore, *FMR1* testing is recommended in all women with confirmed ovarian failure. It is well established that only a subset of PM carriers develop FXPOI (about 13-26%), with the precise molecular mechanisms of how the *FMR1* premutation causes FXPOI not being well understood. The most important findings regarding FXPOI are: the significant positive association of repeat size, so that women with fewer than 100 repeats have an increased risk of FXPOI.

Fragile X Tremor Ataxia Syndrome is a late-onset neuropsychiatric degenerative disorder that occurs in male carriers of the *FMR1* PM. It is estimated that 1 in 3,000 men older than 50 years will present with FXTAS. Clinical symptoms in FXTAS patients usually begin with an action tremor and gradually progress to balance problems with frequent falling, and more variably loss of sensation in the distal lower extremities and autonomic dysfunction. It has been hypothesized that elevated *FMR1* mRNA levels and a toxic gain-of-function could be the molecular mechanism responsible for the syndrome. Since the first description of FXTAS in 2001, our understanding of the clinical phenotype as well as the molecular pathophysiology has advanced very rapidly.

Regarding individuals carrying a PM there is a growing body of literature suggesting that neuropsychiatric features of fragile X-associated tremor/ataxia syndrome (FXTAS) follow a fronto-subcortical pattern with primary impairments in executive function and increased vulnerability to mood and anxiety disorders. Increased rates of psychiatric symptoms may represent early markers of neurodegenerative diseases and have also been reported among *FMR1* PM carriers without FXTAS.

Numerous studies have investigated the neuropsychological phenotypes among PM allele carriers in women with conflicting results. An increased risk of anxiety and mood disorders among PM allele carriers has not yet been established, although they seem to be more common in these subjects than in controls.

Premutation alleles give rise to an increased risk in relation to additional medical and cognitive features which occur at a greater frequency than what would be expected for the general population. These clinical features include peripheral neuropathy, immune-mediated disorders, migraines and neurocognitive involvement.

The continuous advances in the knowledge of the FMR1 gene and its allelic forms have changed the approaches to genetic counseling. Indeed, taking into account the great complexity of the implication of this gene in different pathologies, genetic counseling should always be given by an expert geneticist.

Lastly the treatment of FXS and related pathologies is mainly symptomatic and is addressed to improve or solve some of the more limiting symptoms such us hyperactivity, deficit attention disorder, and behavioral problems or language anomalies. Clinical trials on new therapeutic targets (i.e. mGluR) are currently ongoing.

I hope that the overview of the current knowledge of the *FMR1* gene provided in this book is useful in clinical practice and leads to new lines of research.

In: Allelic Forms of the FMR1 Gene
Editor: Montserrat Milà
ISBN: 978-1-63321-914-4

Chapter 1

Fragile X Syndrome

M. Milà[*1,2,3]
[1]Servicio de Bioquímica i Genética Molecular.
Hospital Clínic, Barcelona, Spain
[2] CIBER de Enfermedades Raras (CIBERER) Barcelona, Spain
[3] IDIBAPS (Institut de Investigacions Biomèdiques August Pi I Sunyer)
Barcelona, Spain

Abstract

Fragile X syndrome (FXS) is the most common cause of familial intellectual disability and the most commonly known single gene causing autism spectrum disorders. The incidence varies according to the populations; it is well-accepted that 1/4,000 males and 1/6,000 females are affected, and 1/250 females and 1/800 males are carriers. The main clinical manifestations are intellectual disability, dimorphic traits and behavior disturbances. The syndrome is inherited as a dominant X-linked trait with reduced penetrance (80% for males and 30% for women). This syndrome is due to a functional loss of the *FMR1* gene product, Fragile X Mental retardation Protein (FMRP), and, in most cases it is caused by a CGG repeat expansion in the *FMR1* promoter. The repeat is from 6 to 55 CGGs long in the normal population while in patients with FXS the repeat number is over 200 CGGs (full mutation FM).This number of CGGs generally leads to methylation of the repeat and the promoter

* Montserrat Milà; mmila@clinic.ub.es.

region, which is accompanied by silencing of the *FMR1* gene. The absence of the FMR1 protein, FMRP, is the cause of the intellectual disability in these patients. Individuals with 55 to 200 CGGs carry a premutation (PM). All affected children have carrier mothers (full mutation [FM] or PM) with a 50% of chance of having another affected child in future pregnancies. Female PM carriers are at risk of developing primary ovarian insufficiency (FXPOI). Elderly PM carriers (males and females) may develop a progressive neurodegenerative disorder called fragile X-associated tremor/ataxia syndrome (FXTAS). The *FMR1* gene is responsible for different disorders depending on the length of the CGG tract and the molecular mechanism. The Fragile X syndrome is due to a loss of function, and FXPOI and FXTAS are due to a toxic gain of function of the mRNA or repeat-associated-non-AUG translation. At present, there is no treatment, although several studies are ongoing in both human and animal models.

The present and the following chapters review the current status of the wide spectrum of different pathologies related to the *FMR1* gene.

Keywords: Fragile X syndrome, *FMR1* gene, CGG expansion, Intellectual disabilities, autism spectrum disorder

Introduction

Intellectual disability (ID) affects 1-3% of the general population; approximately half of the ID cases have a familial origin. On the other hand the autism and autism spectrum disorders (ASD) are estimate to affect 1% of general population. The exact causes of ID and ASD are unknown, although it is thought that several complex genetic and environmental factors are involved. At present in around 50% of the cases with ID or ASD, the genetic defect remains unknown being difficult to give genetic counselling.

The first monogenic cause for these two pathologies is the *FMR1* gene which constitute the most common cause of inherited intellectual disability, developmental delay and the most commonly known single gene causing autism spectrum disorders.

This chapter it is focus on the Fragile X syndrome (FXS) reviewing the different aspects clinical, molecular basis, animal models, genetic counseling, and neonatal screening.

Fragile X Syndrome

Fragile X syndrome (FXS) (FXS #MIM300624; ORPHA 908) is the most common cause of inherited intellectual disability (ID), developmental delay and is the most commonly known single gene causing autism spectrum disorders (ASD).The *FMR1* gene (Fragile X mental retardation type 1 gene) is inherited as a X-linked dominant trait with a reduced penetrance of 80% in males and 30-50% in females.

In 1943 Martin and Bell made the first description of the syndrome in males with ID and affecting more than one male in a family. The most remarkable clinical signs were moderate to severe ID, having a long face, prognatism, large ears, machroorchidism, and alterations in connective tissue as well as characteristic behavior.

The name "fragile X syndrome" was attributed to Lubs, in 1969 [1] who observed fragility at the edge of the log arm of the X chromosome. However, this fragility was only observed in a reduced number of metaphases and it was more evident in cultures in which the media was poor in folic acid.

In 1991 three groups working independently identified the gene responsible and the molecular defect [2-5].Fragile X syndrome was the first human disease to be found to be caused by a dynamic mutation; the main mutation is a CGG expansion in the untranslated region of the first exon of the *FMR1* gene (Fragile X mental retardation type 1).

The incidence is variable depending on the population, although in general an incidence of 1/4,000 males and 1/6,000 females is well accepted, thereby considering FXS as a rare disease [6].

Clinical manifestations also vary depending on the age and sex [7-12] (Figure 1).

Prepuberal males show moderate ID with a delay in the milestones (seated, speech and ambulation) and behavioral disturbances: deficit in attention, hyperactivity, hand biting, autistic behavior, refusal to touch, shyness and gaze avoidance, repetitive movements especially of the hands and echolalia, all of which become more accentuated with age. Although the ID is usually moderate, it is not uncommon to find severe forms with an IQ between 35-55 and psychiatric disorders. In post pubertal males the physical traits are more pronounced, ID is more severe and macroorchidism is present in 80% of the cases.

Figure 1. Phenotypic appearance of two affected brothers with FXS.

The physical traits of FXS are initially quite subtle, and the first warning sign is often a delay in the emergence of language, which can be accompanied by a slight engine, hyperactive behavior, attention deficit, autistic-like behaviors such as flapping and / or hand biting, and poor eye contact. On physical examination, an elongated face with prominent chin and large and protruding ears is observed. Other relevant clinical data include joint hypermobility, especially of small joints, fine and velvety skin, being common to see many wrinkles on the palms of the hands. They usually have a high palate. Regarding the heart, the most common finding is a prolapse of the mitral valve that manifests as a mild to moderate heart murmur which is usually asymptomatic. These and other anomalies are considered mainly due to an alteration of the connective tissue.

Individuals with FXS often have no serious medical problems. Large studies have reported wide discrepancies in the prevalence of some associated medical problems. This fact is probably due to the great variability in these patients. The most prevalent medical problems are: the mitral valve prolapse, recurrent otitis media, seizures, motor tics, strabismus, sleep problems and obstructive sleep apnea. It is known that during childhood otitis (recurrent otitis media), and subsequently, sinusitis are frequent. Convergent strabismus is also often present. Many children have flat feet, but this finding tends to improve with age. Approximately 30% of patients have some degree of gastroesophageal reflux during the first year, the treatment of which depends on the severity of the reflux. In the neurological area about 15% of patients with FXS epilepsy have some abnormal EEG findings. In a subgroup of patients with FXS the syndrome is similar to the Prader-Willi syndrome with obesity and hyperphagia phenotype.

The speech of these subjects is often repetitive; they behave shyly and look away when spoken to. In addition, they show tactile defensiveness, gaze avoidance, poor concentration, impulsivity and they sometimes present aggressive behavior. A behavioral phenotype of autism is common in males. It is usual to present stereotyped movements with both hands (shaking), self-harming biting that causes callused areas on the hands and they also show defensiveness or over-reaction to certain external sensory stimuli. Poor social skills and socio-emotional reciprocity can be added to this sometimes progressive deterioration and this can lead to a diagnosis of an autism spectrum disorder.

In general girls show attenuated clinical manifestations; mild ID and phenotypic traits are missing or minimal (big ears, joint hypermobility, elongated facies), although some cases presenting similar clinical manifestations to those of males have been reported. From 30-50% of women have ID and most have learning disabilities, behavioral disorders and communicate with shyness, presenting social anxiety, attention problems and educational delay, especially in mathematics.

The clinical diagnosis may be very difficult in girls and women due to the variable emotional and behavioral characteristics and the subtle physical traits [7-11].

Approximately 30–50% of FXS patients meet full DSM-IV-TR criteria for autism, with 60–74% fulfilling criteria for an autism spectrum disorder Over 90% of individuals with FXS display some form of atypical behavior characteristics of autism including social interaction (e.g. avoidance of eye contact, social withdrawal, social anxiety) and repetitive and stereotyped behaviors [12-13]

The lack of function of the *FMR1* gene is the cause of FXS. This gene is located in the long arm of the X chromosome at Xq27.3, expands 17 exons and 40 kb of genomic DNA. It transcribes an mRNA of 3.9 kb. The translated protein is called *Fragile X Mental retardation Protein* (FMRP), which is necessary from early stages of development and throughout life [14].

In most cases the molecular basis of the syndrome is a dynamic mutation: an expansion and hypermethylation of an unstable CGG trinucleotide repeat located in the first exon of the *FMR1* gene [2-4]. In the general population the number of CGG repeats is polymorphic, presenting from 6 to 54 CGGs, and the adjacent CpG island, which acts as a promoter is non-methylated. This CpG island is located in the 5'untranslated region of the gene and acts as a switch depending on its methylation status. When the gene is active, *FMR1* is transcribed and translated. Alleles harboring 6 to54 CGGs are considered normal and remain stable upon transmission.

A second class of alleles that overlaps with the upper normal range contains 45 to 54 CGGs. This range, known as the gray zone, can be stable or slightly unstable, being transmitted to subsequent generations with the possibility of expanding to a premutated allele. The clinical implications of this class of alleles remain unclear [15-16]. The next class is premutated alleles with a range of 55 to about 200 CGG repeats. In this situation, the *FMR1* gene is also transcribed and translated because the CpG island is non-methylated. Therefore, premutated carriers have normal or slightly reduced synthesis of FMRP and increased levels of mRNA (2-8 fold more than normal alleles) and they are asymptomatic for FXS. However, these carriers have a risk of having affected offspring since the number of CGG is unstable and tends to increase the CGG number in each cellular division.

People carrying permutated alleles are at risk of developing some disorders characteristic of this status such as: premature ovarian insufficiency (FXPOI), Fragile X tremor ataxia syndrome (FXTAS) and emotional disturbances, among others (see Chapters 2 and 3).

Finally, when the CGG number is over 200 repeats, these alleles are in the FM range. The CpG island and the repeats themselves are methylated and this methylation switches off the gene, blocking transcription and no protein is translated. These individuals are always affected with FXS if they are males, with only 30-50 % of females being affected (Figure 2). Moreover, around 15-25 % of patients show a pattern, referring to a mixture of PM and FM, which is caused by the somatic instability of the FM in early embryogenesis that can lead to retraction of the expanded repeat. Another kind of mosaicism can be observed in large expansions with incomplete methylation. The mosaicisms allow the expression of some FMRP and, in some cases, have been associated with milder ID in males.

Full Scale IQ scores were inversely correlated with the percentage of methylation and positively correlated with higher FMRP expression. These latter results point toward a positive impact on cognition for FM mosaics, with lower methylation compared to fully methylated individuals. Recent studies have suggested that low expression of FMRP may be sufficient to have a positive impact on cognitive function in these individuals. [17]. Individuals carrying the FM exhibit the classical fragile X phenotype that includes the broad spectrum of clinical manifestations described previously in this chapter and never present premutation-associated disorders such as FXPOI and FXTAS.

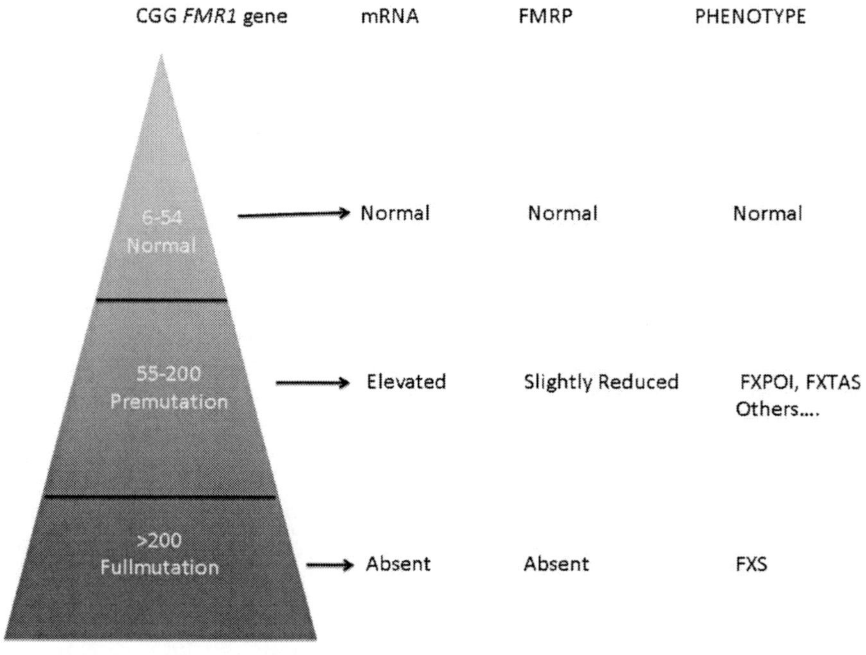

Figure 2. Summary of molecular aspects of the FMR1 gene.

It is well known that some AGG are interspersed within the CGG expansion in the *FMR1* gene. The biological function of these interruptions seems to stabilize the gene during transmission and decreases the risk of DNA polymerase slippage during DNA replication. Previous studies have reported important data such as: AGG interruptions influence the stability of the CGG repeats within the *FMR1* gene during parental transmission, the presence or absence of AGG interruptions is not correlated with the transcriptional or translational activity of the gene, and AGG interruption patterns can vary greatly between populations, but are largely inherited without change [18]

In most cases the FXS is due to a CGG repeat expansion in the *FMR1* promoter, but other *FMR1* mutations leading to a loss of function of the gene may also cause FXS or an FXS-like phenotype. Since standard molecular testing does not include sequencing of the *FMR1* coding region, the prevalence of point mutations causing FXS is not well known, although it seems that missense mutations in the *FMR1*gene might account for a considerable proportion of cases in males with FXS-related symptoms, such as those linked to ID and developmental delay. At present, the estimated frequency of point mutations is 1 to 2% of the cases of FXS [19].

The diagnosis of FXS is achieved by molecular analysis that determines the precise number of CGG in the *FMR1* gene [20]. Cytogenetic studies are currently not accepted as a diagnostic test. The quickest diagnostic method is polymerase chain reaction (PCR) using fluorescent labeled primers and subsequent analysis of the product in an image analyzer to determine the exact number of CGG repeats using small quantities of DNA. In males, the absence of PCR product indicates the presence of a pathological expanded allele.

The limitation of this technique is that it does not detect expansions of over 100 repeats and therefore does not detect large PMs or FMs and provides no information about methylation. Another limitation of PCR is that it does not distinguish between a homozygous female (both chromosomes harboring the same CGG number) and a woman with an allele in the normal range and another allele in the large PM range. These problems may be solved by conducting another PCR called Triplet-PCR (TP-PCR) which is able to identify large alleles and to distinguish between a homozygous female and a female carrying pre or full mutated alleles. Another specific PCR is necessary to detect methylation status [21-23].

These techniques are used for prenatal and postnatal diagnosis. DNA obtained from any tissue including saliva, blood, amniotic fluid, or chorionic villi can be used, thus, the choice of technique and tissue depends on the diagnostic strategy based on the clinical and family characteristics.

When an individual harboring an expansion is identified, a cascade family study is required to detect other relatives carrying pre or full mutations. In these cases precise determination of allele size and the presence of the AGG interruption are important to determine the risks of expansion in the carriers. Moreover, the status of methylation or the presence of mosaicism is relevant for genotype-phenotype correlations [24].

Finally, a test based on immunohistochemical methodology using a monoclonal antibody against the FMRP is used to detect the presence of this protein. This study may be useful in large-scale screening of the male population since males with a FM have no or a very low expression of FMRP. This test can also be useful in doubtful cases with a high PM phenotype suggestive of FXS. However, it is not useful in females, since the presence of the second normal X chromosome could modify the expression. [25-26]

In 1991, when the *FMR1* gene was cloned, Southern-blot analysis was the gold standard technique, but it is a laborious technique that requires several days of work, large amounts of DNA and gives imprecise estimates of the number of repetitions.

At present, the use of diagnostic commercial kits provided by several companies simplifies the work and ensures a safe diagnosis replacing the old techniques (Figure 3).

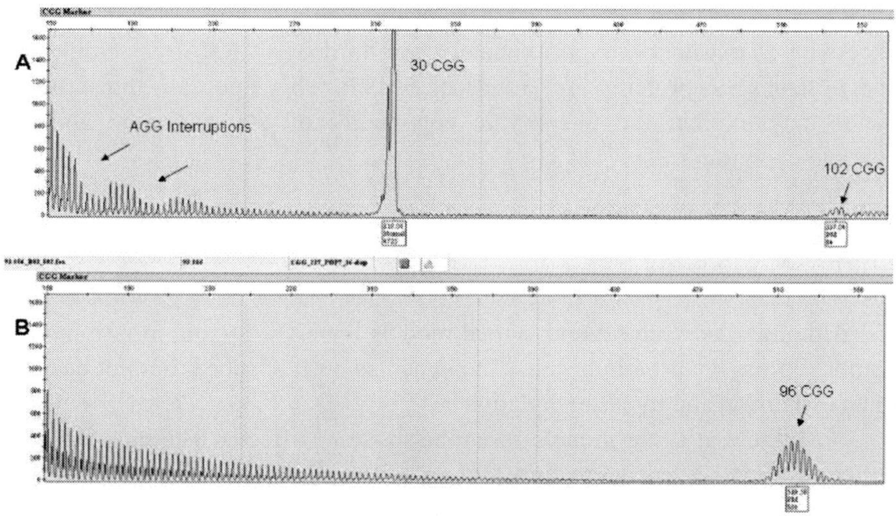

Figure 3. Electropherogram obtained using Amplidex Fragile X kit Asuragen. A) product from a female carrying one allele 30 CGG in normal range and the other allele with 102 CGG in permutation range, the arrows show the position of AGG interruptions and the allele size. B) Electropherogram corresponding to a PM male with 96 CGG and no AGG interruptions, were detected.

FMRP (Fragile X Mental Retardation Protein)

Cognitive impairment is caused by the absence of the FMRP in neurons. FMRP expression is widespread in neurons and spermatogonia. It is ubiquitously expressed from the early stages of development throughout postnatal life. The localization of FMRP is largely cytoplasmatic, being associated with the polyribosomes attached to the endoplasmatic reticulum and with free ribosomes at the bases of dendrites and within dendritic spines [27] FMRP plays a fundamental role in the synapses and normal development of dendrites. In healthy neurons, FMRP modulates the local translation of numerous synaptic proteins; synthesis of these proteins is required for the maintenance and regulation of long-lasting changes in synaptic strength. FMRP exerts profound effects on synaptic plasticity [28].

The *FMR1* has 17 exons with alternative splicing in exons 12, 14, 15 and 17 resulting in the expression of multiple isoforms of FMRP. The distribution of these isoforms is different according to the different brain regions except for the hippocampus and the olfactory bulb. FMRP binds to approximately 4% of all RNAs, regulating protein synthesis; and the lack of FMRP implies an excess basal translation. Approximately one third of all RNAs encoding pre- and postsynaptic proteins are targets of FMRP. This role as a transcription factor may explain the phenotypic complexity of the syndrome and the variable expression.

Animal Models for FXS

Although there are several animal models for FXS, fly and mouse models do not express the protein because the gene has been silenced, but not because of an expansion and hypermethylation.

Nevertheless, these models resemble those of affected patients, showing deficits in spatial learning, defect of prepulse inhibition of acoustic startle response, increased locomotor activity and are more susceptible to having epileptic seizures. These animal models allow the possibility of studying synaptic plasticity in many brain areas and the possibility of performing drugs trials [30-32].

For many years it was thought that PM carriers showed no clinical manifestations, only the risk of transmitting the disease to their offspring. In 1999 the presence of a PM became associated with in FXPOI women, and in 2000 a risk of late onset neurodegenerative disease, Fragile X Tremor Ataxia Syndrome (FXTAS), was associated with male and female PM carriers. Lastly, in the years 2007-2009 a range of phenotypes associated with the PM were described. It is now accepted that FXPOI (MIM#311360) and FXTAS (MIM#300623) are different from FXS and are only associated with the *FMR1* gene PM [33-37] (see Chapters 2-5).

All these associated disorders will be discussed in the next chapters. Nevertheless, it is important to highlight the importance of the PM carriers: It is estimated that more than 1 million people in the USA are *FMR1* carriers, with 20% of these subjects developing FXPOI and 40% of males and 16 % of women developing FXTAS. Furthermore, 20% may present cognitive problems such as ASD (8%) and ADHD (30%).

Genetic Counseling

Although genetic counseling is discussed in detail in chapter 6, the following are the main features of this syndrome.

FXS is a genetic disorder that is inherited as a dominant X-linked trait with reduced penetrance (80% for males and 30% for women). In the general population we can find either men or women with alleles in the normal premutated or full mutated range.

Importantly, there are no sporadic cases so that whenever there is an affected child, the mother is an obligated carrier with a 50% of chance of having another affected child in future pregnancies. Members of the family in which the expansion is discarded are not at risk of presenting FXS.

Male carriers of either a PM or a FM always transmit the PM to their daughters, and thus, all carry the PM. The size of the PM is not exactly the same as the parent, and may slightly increase or decrease. Moreover, male carriers never transmit the FM. This is because only a PM and never a FM is found in the sperm. Males carrying a FM expansion transmit this to their daughters while always decreasing the CGG number to the PM range.

In conclusion, since sons inherit the Y chromosome and girls inherit the PM, male offspring are never affected by FXS.

Female carriers transmit the expanded allele to 50% of their offspring, both sons and daughters. Unlike male carriers, women tend to increase the size of the expansions in the passage from one generation to the next, so that a FM is only maternally inherited. The risk of expansion from the PM to the FM in women depends on the size of the PM and the AGG interruptions. Thus, uninterrupted and larger alleles are more likely to become a FM [16]. Indeed, whenever an allele with 100 CGG repeats is transmitted an expansion to full mutation is generated in the next generation. Women with a FM will transmit this allele mutated to 50% of their offspring (sons and daughters), so that their offspring have a 50% risk of being affected with FXS.

It is important to highlight that despite the incomplete penetrance of the full mutated alleles in females only 50% show clinical manifestations. This fact is very important for genetic counseling especially in cases of prenatal diagnosis, when a female fetus carrying the FM is diagnosed. Nonetheless, we cannot know their phenotype.

A female carrier has different reproductive options: no offspring (adoption), ovodonation, prenatal diagnosis (amniotic fluid, corion villi or cord samples), preimplantation diagnosis or preconceptional diagnosis. (all these options will be discussed in chapter 6).

Genetic counseling in carriers of the PM should include the risk of developing pathologies associated with the PM: PM women have a 20% risk of developing a FXPOI and developing menopause before the age of 40.

PM males and females have a risk of developing FXTAS at over 50 years of age, the penetrance is 40% for males and 16% for females. [reviewed in Chapter 3, 24, 38]

At present it is advisable to request a *FMR1* test in the following cases:

1) Boys and girls with ID and autism, since the phenotypic characteristics are subtle in infancy it is recommended to rule out FXS in all cases of ID in whom the etiology is unclear, including a wide range of mild to profound ID, as well as developmental delays, autism, hyperactivity and other behavioral problems.
2) Women with infertility and / or ovarian failure before the age of 40 years, especially in cases with high levels of follicle-stimulating hormone (FSH) and if no other cause is confirmed such as ovarian cancer radiation treatment or thyroiditis.
3) Men and women with tremor and ataxia. Genetic testing of the *FMR1* gene is recommended in cases of cerebral ataxia with parkinsonism and intention tremor of unknown cause and cognitive decline in a person of 50 years of age.
4) Relatives of a diagnosed individual harboring an expansion in the *FMR1* gene. It is also necessary to perform a diagnostic "cascade", ie family studies arising from a former affection, including prenatal diagnosis in cases in which the pregnant woman is a PM or FM carrier.
5) Men and women who are ovum or sperm donors, due the high incidence of the PM in the general population.

Newborn Screening (NBS)

The incidence of this syndrome and its significance require its inclusion in screening programs, however, FXS is currently not included in NBS panels in any country due to its controversial particularities.

First of all, differences are observed in the diagnosis of males and females. For many years no test has been available for screening of females because of technical limitations. Secondly, there is no curative treatment for this syndrome, and finally, the screening could lead to a presymptomatic diagnosis of a neurodegenerative disease such FXTAS.

At present, the development of efficient methodologies makes FXS screening feasible for males and females. Indeed, it is possible to perform large population studies spanning the entire spectrum of *FMR1* mutation with a simple PCR.

Although no curative treatment is available, arguments in favor of NBS include the benefits of early detection that allows early intervention with good results and the possibility of cascade screening in these families and providing genetic counseling to all members.

Finally, a large number of clinical trials have been carried out regarding pharmacological interventions and new-targeted treatments will continue to be developed.

The main controversial aspect is that NBS allow the identification of PM carriers. An ever expanding number of medical disorders, although with a reduced penetrance, can occur in these individuals: ASD, ADHD, FXPOI, FXTAS, seizures, hypertension, migraines, sleep apnea, immune-mediated problems, thereby involving presymptomatic diagnosis of some diseases in very early stages of life.

This raises ethical problems which are difficult to solve. Ideally NBS for FXS should be included in NBS programs, however some ethical issues continue to make acceptance by ethical committees difficult [39].

Treatment

Although at present there is no treatment for all of these pathologies, Chapter 7 provides an update of the clinical trials on therapies for all these *FMR1* gene-related disorders and their current status.

The description of new specific treatment for experimental animal models of FXS involving therapeutic targets has allowed the discovery of drugs that are being tested and validated for treatment in humans.

The tests are intended to determine whether any of the compounds can attenuate or even normalize the clinical symptoms in patients with FXS. Until these drugs are authorized symptomatic treatments are able to control part of the disease and should be individually studied in each case to assess their effectiveness in controlling the symptoms present in each patient [40].

Conclusion

Fragile X syndrome is the most common cause of inherited ID, developmental delay and is a monogenic cause of autism spectrum disorders. The *FMR1* gene is responsible for different disorders: FXS (loss of function), FXPOI and FXTAS (toxic gain of function or repeat-associated-non-AUG). Although no treatment is currently available, early detection facilitates early intervention with good results and the possibility of performing cascade screening in these families and providing genetic counseling to all members.

Table 1. Summary of the different status of the *FMR1* gene, risk of transmission and risk for the associated pathologies

CGG number	Gender	Status	Risk affected offspring	Risk develop FXPOI, FXTAS
6-45	male	Normal	No	No
	female	Normal	No	No
46-54	male	Normal	No	low
	female	Normal	No	low
55-200	male	Carrier	All daughters carrying premutation	40% FXTAS
	female	Carrier	50% FXS Depending of CGG and AGG number	20% FXPOI 40% FXTAS
>200	male	Affected FXS	All daugthers carrying premutation	No
	female	30-50% Affected FXS	50% FXS	No

References

[1] Lubs HA. A marker X chromosome. *Am J Hum Genet.* (1969) May;21(3):231-44.

[2] Oberlé I, Rousseau F, Heitz D, Kretz C, Devys D, Hanauer A, Boué J, Bertheas MF, Mandel JL. Instability of a 550-base pair DNA segment

and abnormal methylation in fragile X syndrome. *Science.* 1991 May 24;252(5009):1097-102.

[3] Verkerk AJ, Pieretti M, Sutcliffe JS, Fu YH, Kuhl DP, Pizzuti A, et al. Identification of a gene (FMR-1) containing a CGG repeat coincident with a breakpoint cluster region exhibiting length variation in fragile X syndrome (1991). *Cell* 65:905-14.

[4] Yu S, Pritchard M, Kremer E, Lynch M, Nancarrow J, Baker E et al. (1991). Fragile X genotype characterized by an unstable region of DNA. *Science* 252: 1179-81

[5] Garber KB, Visootsak J, Warren ST. Fragile X Syndrome. *Eur J Hum Genet* 2008; 16:666-672.

[6] Hagerman RJ. The physical and behavioural phenotype, in: RJ Hagerman, P Hagerman (Eds.), Fragile-X Syndrome: Diagnosis, Treatment and Research, *The Johns Hopkins University Press, Baltimore*, 2002, pp. 3–109

[7] Farzin F, Perry H, Hessl D, Loesch D, Cohen J, Bacalman S, et al. Autism spectrum disorders and attention-deficit/hyperactivity disorder in boys with the fragile X premutation. *J DevBehavPediatr 2006;* 27:S137–144.

[8] Kidd SA1, Lachiewicz A2, Barbouth D3, Blitz RK4, Delahunty C5, McBrien D6, Visootsak J7, Berry-Kravis E8. Fragile X Syndrome: A Review of Associated Medical Problems Pediatrics. 2014 Oct 6.pii: peds.2013-4301.

[9] Loesch D1, Hagerman R. Unstable mutations in the FMR1 gene and the phenotypes. *Adv Exp Med Biol.* 2012;769:78-114.

[10] Turkstra LS, Abbeduto L, Meulenb *Am J Intellect DevDisabil.* Social cognition in adolescent girls with fragile X syndrome. 2014 Jul;119(4):319-39.

[11] Schneider A, Ligsay A, Hagerman RJ. Fragile X syndrome: an aging perspective. *Dev Disabil Res Rev*. 2013 Aug;18(1):68-74

[12] McCary LM, Roberts JE.Early identification of autism in fragile X syndrome: a review. *J Intellect Disabil Res*. 2013 Sep;57(9):803-14. Review

[13] Bailey DB, Raspa M, Holiday D, Bishop E, Olmsted M. Functional skills of individuals with fragile x syndrome: a lifespan cross-sectional analysis. *Am J Intellect Dev Disabil.* 2009 Jul;114(4):289-303

[14] Pieretti M, Zhang FP, Fu YH, Warren ST, Oostra BA, Caskey CT, Nelson DL. Absence of expression of the FMR-1 gene in fragile X syndrome. *Cell.* 1991 Aug 23;66(4):817-22.

[15] Madrigal I, Xunclà M, Tejada MI, Martínez F, Fernández-Carvajal I, Pérez-Jurado LA et al (2011). Intermediate FMR1 alleles and cognitive and/or behavioural phenotypes. *Eur J Hum Genet* 19: 921-3.

[16] Nolin SL, Sah S, Glicksman A, Sherman SL, Allen E, Berry-Kravis E, Tassone F, Yrigollen C, Cronister A, Jodah M, Ersalesi N, Dobkin C, Brown WT, Shroff R, Latham GJ, Hadd AG. Fragile X AGG analysis provides new risk predictions for 45-69 repeat alleles. *Am J Med Genet A*. 2013 Apr;161A(4):771-8. doi: 10.1002/ajmg.a.35833. Epub 2013 Feb 26.

[17] Pretto D, Yrigollen CM, Tang HT, Williamson J, Espinal G, Iwahashi CK, Durbin-Johnson B, Hagerman RJ, Hagerman PJ, Tassone F. Clinical and molecular implications of mosaicism in FMR1 full mutations. *Front Genet.* 2014 Sep 17;5:318.

[18] Yrigollen CM1, Martorell L2, Durbin-Johnson B3, Naudo M2, Genoves J2, Murgia A4, Polli R4, Zhou L5, Barbouth D6, Rupchock A6, Finucane B7, Latham GJ8, Hadd A8, Berry-Kravis E5, Tassone F9. AGG interruptions and maternal age affect FMR1 CGG repeat allele stability during transmission. *J NeurodevDisord*. 2014;6(1):24.

[19] Handt M, Epplen A, Hoffjan S, Mese K, Epplen JT, Dekomien G Point mutation frequency in the FMR1 gene as revealed by fragile X syndrome screening. *Mol Cell Probes.* 2014 Aug 27 S0890-8508(14)00042-5.

[20] Biancalana V, Glaeser D, McQuaid S, Steinbach P. EMQN best practice guidelines for the molecular genetic testing and reporting of fragile X syndrome and other fragile X-associated disorders. *Eur J Hum Genet.* 2014 Sep 17.doi: 10.1038/ejhg.2014.185.

[21] Filipovic-Sadic S, Sah S, Chen L, Krosting J, Sekinger E, Zhang W, Hagerman PJ, Stenzel TT, Hadd AG, Latham GJ, Tassone F. A novel FMR1 PCR method for the routine detection of low abundance expanded alleles and full mutations in fragile X syndrome. *Clin Chem.* 2010 Mar;56(3):399-408.

[22] Hersh JH, Saul RA; Committee on Genetics. Health supervision for children with fragile X syndrome. *Pediatrics* 2011;127:994-1006.

[23] Weck KE, Zehnbauer B, Datto M, Schrijver I; CAP/ACMG Biochemical and Molecular Genetics Resource Committee. Molecular genetic testing for fragile X syndrome: laboratory performance on the College of American Pathologists proficiency surveys (2001-2009). *Genet Med* 2012;14:306-12.

[24] McConkie-Rosell A, Abrams L, Finucane B, Cronister A, Gane LW, Coffey SM, et al. Recommendations from multi-disciplinary focus groups on cascade testing and genetic counseling for fragile X-associated disorders. *J Genet Couns* 2007;16:593-606.

[25] Castellví-Bel S, Fernández-Burriel M, Rifé M, Jiménez D, Mallolas J, Sánchez A, et al. Detection of the fragile X syndrome protein for the evaluation of FMR1 intermediate alleles. *Hum Genet.* 2000 ;107:195-6.

[26] Ramos FJ, Willemsen R. Diagnosis of the fragile X syndrome by the analysis of FMRP expression in blood and hair roots *Arch Pediatr.* 2003;10:401-2

[27] Oostra BA1, Willemsen R. FMR1: a gene with three faces. *Biochim Biophys Acta.* 2009 Jun;1790(6):467-77.

[28] Sidorov MS1, Auerbach BD, Bear MF. Fragile X mental retardation protein and synaptic plasticity. *Mol Brain.* 2013 Apr 8;6:15. doi: 10.1186/1756-6606-6-15.

[29] Willemsen R, Levenga J, Oostra BA. CGG repeat in the FMR1 gene: sizematters. *Clin Genet.* 2011 Sep;80(3):214-25

[30] Levenga J, de Vrij FM, Oostra BA, Willemsen R. Potential therapeutic interventions for fragile X syndrome. *Trends Mol Med.* 2010 Nov;16(11):516-27.

[31] Hunsaker MR, Goodrich-Hunsaker NJ, Willemsen R, Berman RF. Temporal ordering deficits in female CGG KI mice heterozygous for the fragile X premutation. *Behav Brain Res.* 2010 Dec 1;213(2):263-8.

[32] Santos AR, Kanellopoulos AK, Bagni C. Learning and behavioral deficits associated with the absence of the fragile X mental retardation protein: what a fly and mouse model can teach us. *Learn Mem.* 2014 Sep 16;21(10):543-55.

[33] Sullivan SD, Welt C, Sherman S. FMR1 and the continuum of primary ovarian insufficiency. *Semin Reprod Med.* 2011 Jul;29(4):299-307 Review.

[34] Wheeler AC, Raspa M, Green A, Bishop E, Bann C, Edwards A, Bailey DB Jr. Health and reproductive experiences of women with an FMR1premutation with and without fragile X premature ovarian insufficiency. *Front Genet.* 2014 Sep 8;5:300.

[35] Rodriguez-Revenga L, Madrigal I, Pagonabarraga J, Xunclà M, Badenas C, Kulisevsky J, Gomez B, Milà M. Penetrance of FMR1 premutation associated pathologies in fragile X syndrome families. *Eur J Hum Genet.* 2009 Oct;17(10):1359-62

[36] Hagerman R1, Hagerman P.Lancet Neurol. 2013 Aug;12(8):786-98. Advances in clinical and molecular understanding of the FMR1premutation and fragile X-associated tremor/ataxia syndrome.

[37] Tassone F, Greco CM, Hunsaker MR, Seritan AL, Berman RF, Gane LW, Jacquemont S, Basuta K, Jin LW, Hagerman PJ, Hagerman RJ.Neuropathological, clinical and molecular pathology in female fragile X premutation carriers with and without FXTAS. *Genes Brain Behav.* 2012 Jul;11(5):577-585.

[38] New research on Fragile X syndrome Jeffrey J Upner (editor) *Nova Science Publishers Inc* NY ISBN 13 978-1-60021-281-9 Chapter 6 97-117 L Rodriguez-Revenga and Montserrat Milà

[39] Tassone F. Methylation analysis in newborn screening for fragile X syndrome--reply. *JAMA Neurol.* 2014 Jun;71(6):800-1. doi: 10.1001/jamaneurol.2014.45430.

[40] Hagerman RJ, Des-Portes V, Gasparini F, Jacquemont S, Gomez-Mancilla B. Translating molecular advances in fragile X syndrome into therapy: a review. *J Clin Psychiatry.* 2014 Apr;75(4):e294-307. doi: 10.4088/JCP.13r08714. Review.

In: Allelic Forms of the FMR1 Gene
Editor: Montserrat Milà
ISBN: 978-1-63321-914-4

Chapter 2

Fragile X-Associated Primary Ovarian Insufficiency (FXPOI)

Maria-Isabel Tejada
Laboratorio de Genética Molecular, Servicio de Genética,
Hospital Universitario Cruces,
BioCruces Health Research Institute, Barakaldo-Bizkaia,
GCV-CIBER de Enfermedades Raras (CIBERER), Spain

Abstract

Fragile X-associated primary ovarian insufficiency (FXPOI) is among the family of disorders caused by the expansion of a CGG triplet repeat in the *FMR1* gene. FXPOI is a new clinical entity in which, carrier premutation (PM) females (with 56 to 200 CGG repeats) present early ovarian dysfunction, with menopause occurring 5 years earlier than non-carrier family members.

It has been estimated that 2-6% of all women with premature ovarian failure and a normal karyotype have a PM. Therefore, *FMR1* testing is recommended in all women with confirmed ovarian failure; i.e. cessation of the menstrual cycle during three-four months with elevated FSH levels of >30U/L.

Despite abundant literature, all authors agree that only a subset of PM carriers develop FXPOI (about 13-26%), with the precise molecular mechanisms of how the *FMR1* premutation causes FXPOI not being well understood. Nonetheless, recent studies have attempted to provide some insight into these mechanisms. This chapter summarizes some of these studies. Among these reports, the most important findings are: 1) the significant positive association of repeat size with POI, demonstrating that women with fewer than 100 repeats have an increased risk of FXPOI, and 2) the hypothesis that the *FMR1* PM may have a toxic RNA gain-of-function effect on ovarian follicle dynamics. These findings have also been demonstrated in rodent models in which the *FMR1* gene protein (FMRP) is highly expressed in oocytes which are important for folliculogenesis. Indeed, the two PM mouse models studied to date have shown evidence of ovarian dysfunction and increased expression of *FMR1* mRNA in the ovary.

Keywords: FMR1, FXPOI, ovarian dysfunction, POF

Introduction

Fragile X-associated primary ovarian insufficiency (FXPOI) is among the family of disorders caused by the expansion of a CGG triplet repeat in the *FMR1* gene. The most important syndrome caused by this expansion is Fragile X syndrome (FXS).

As seen in Chapter 1, FXS is the most common cause of hereditary intellectual disability (ID) and is one of the most frequent genetic diseases [1]. It is a dominant X-linked disease with incomplete penetrance, affecting approximately one in 2633 men in Spain [2]. This syndrome is caused by the anomalous expansion of a CGG repeat sequence in the 5' untranslated region (UTR) of the X-linked gene *FMR1* located in the FRAXA locus in Xq27.3 [3]. Transmission of this syndrome is complex since the number of CGG repeats in the population appears in four ranges: normal (N), with repeats between 6 and 45; intermediate between 46 and 54; premutation (PM) between 55 and 200; and full mutations (FM) with over 200 repeats (Figure 1). In the latter case, a second mechanism is triggered; that is, the methylation of the *FMR1* gene promoter, resulting in a silencing of thereby not allowing the production of the Fragile X Mental Retardation Protein (FMRP) which is the real cause of the syndrome [4]

Women with a PM do not generally have ID since the FMR1 gene is not methylated (Figure 2). Nonetheless, these women have the risk of transmitting the syndrome [3], with PMs expanding to a FM in successive generations. Additionally, intermediate or grey alleles may or may not be unstable [5], with any expansion to a FM until successive generations.

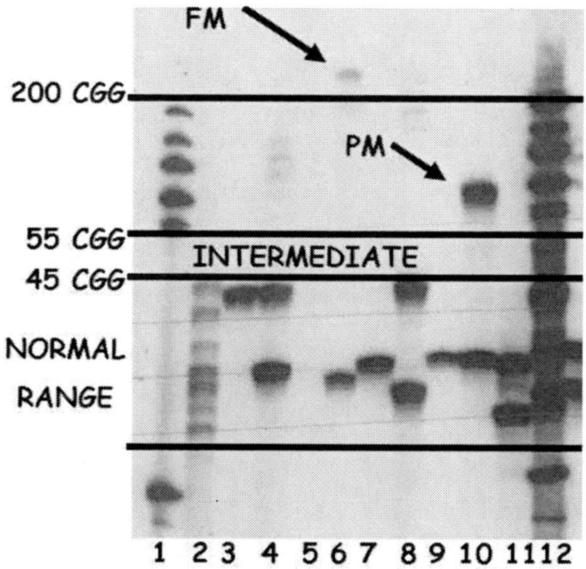

Figure 1. PCR Study of CGG repeats in females. PCR products analyzed by electrophoresis through a denaturing acrylamide gel, blotting, hybridization with an oligo (CGG)5 and visualized by autoradiography after detection with Digoxigenin and CSPD. Lanes 1 and 12: molecular DNA weight markers. Lane 2: in home ladder. Lanes 4, 8and 11: heterozygous normal females. Lanes 3, 7 and 9: homozygous normal females. Lane 6: female with the FM and lane 10: female with the PM. (Figure from the laboratory of Dr. Tejada).

Patients with a PM were initially thought to be completely asymptomatic, however over time this belief has changed. Fragile X families have been studied since 1996 [6], showing a higher frequency of premature ovarian failure (POF) compared with the normal population. In 1999, an international collaborative study [7] reported that 16% of the carriers with a PM had POF or presented early menopause (before the age 40). Since then many groups studying FXS have confirmed this disorder in PM carriers [8-10].

Over the years POF resulting in early menopause has been confirmed as a phenotypic characteristic of PM carriers although only about 13-26% of PM carriers actually present this disorder (1:650 females) [11]. On the other hand, an estimated 2-6% of all women with POF and a normal karyotype have a PM [8,12]. Interestingly, FM carriers do not present ovarian dysfunction. To the contrary, the loss of normal ovarian function has been demonstrated in women with intermediate alleles [11,13], and it has even been suggested that women with alleles with more than 30 repeats have a higher risk of having a diminished ovarian reserve [14].

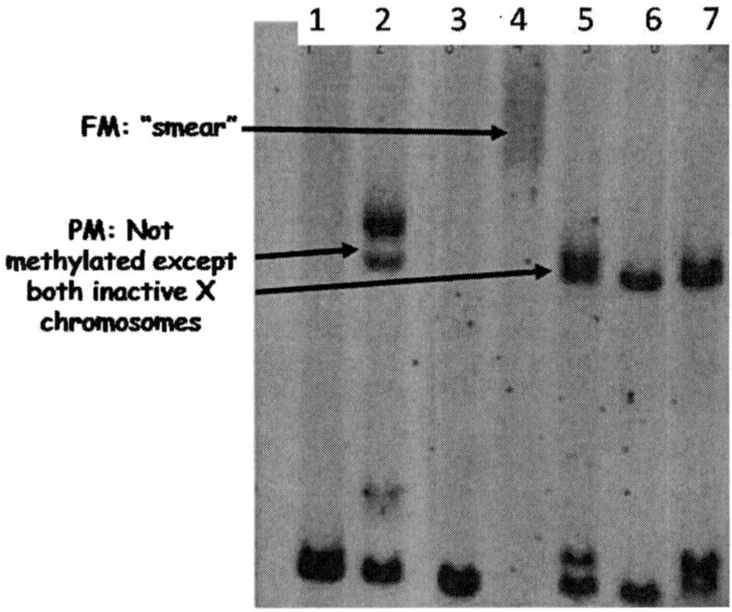

Figure 2. Southern blot analysis. DNA digested with EcoRI and EagI. Hybridization with the StB12.3 probe visualised by autoradiography after detection with Digoxigenin and CSPD. Lanes 1 and 3: normal males. Lanes 2, 5 and 7: females with the PM. Lanes 4: an affected FM male and lane 6: normal female. (Figure from the laboratory of Dr. Tejada).

In 2007, the National Fragile X Foundation defined the term "FXPOI" [15] as a condition in which there is a primary ovarian insufficiency (POI) with menopause occurring at 40 to 45 years of age in association with a PM of the *FMR1* gene [16]. The term –FXPOI- was considered more appropriate than Premature Ovarian Failure (FXPOF). Indeed from the point of view of reproductive counselling, women with FXPOI are able to become pregnant following periods of ovarian failure while those with POF are not. Nonetheless, these pregnancies may lead to a child with FXS.

Defining the Disorder

Primary ovarian insufficiency is also known as premature menopause, premature ovarian failure, hypergonadotropic amenorrhea and hypergonadotropic hypogonadism, conditions occurring due to chemotherapy, radiation or surgery [17]. At present, the term POI is increasingly used because it better describes impaired ovarian function as a continuum rather than a specific endpoint [18]. FXPOI may be transient or progressive, and usually results in eventual premature menopause. Menopause is the permanent cessation of menses that normally occurs at an average age of 50 years, whereas POF and POI generally describe a disorder consisting of amenorrhea, elevated menopausal level gonadotropins and sex steroid deficiency in women less than 40 years old.

In 2004, Welt and colleagues [19] reported that PM carriers have early ovarian aging. Since then, many different publications have appeared [reviewed in 16,20] demonstrating that FXPOI is the result of low ovarian reserves, which is manifested as a decrease in fertility. On comparison with normal control women FXPOI is associated with increased levels of some hormones, especially the follicle stimulating hormone (FSH) and with the decreased duration of one of the menstrual cycle phases, the so-called follicular phase.

Similarly, Anti-Müllerian hormone (AMH) levels, a marker of the growing cohort of small follicles, have been demonstrated to be lower among PM carriers compared with non-carriers of all ages (18 to 50 years). Overall, in women with a PM it seems that AMH is a good indicator of ovarian senescence, being even better than FSH concentrations [16, 20].

In short, FXPOI is a new clinical entity in which PM females (with 55 to 200 CGG repeats) present early ovarian dysfunction, with menopause occurring 5 years earlier than non-carrier family members.

Diagnosis and Genetic Counselling

FMR1 testing is recommended in all women with confirmed ovarian failure: cessation of menstrual cycles during three-four months with elevated FSH (>30U/L) in two determinations carried out within a two-month period and with lower than normal AMH levels. Unfortunately, standardized diagnostic criteria for POI have not yet been established. Nonetheless, determination of estradiol (E2) and prolactine levels in serum may help in establishing the diagnosis [21]. The screening history should also focus on a family history of early menopause, autism and ID. However, in cases in which the diagnosis of ovarian failure is not clear (women with fertility concerns but normal or erratic cycles and women with occult forms of POI) but which present elevated FSH levels, *FMR1*testing is also recommended [20].

This recommendation is especially important in women following fertility treatments since carriers of a PM may have children with FXS [22].

In conclusion, *FMR1* testing is undoubtedly important in women with reduced fertility from the point of view of Genetic Counseling. Nonetheless, CGG studies are also indicated in other women, such as in the case of egg donors due to the high frequency of PMs and intermediate alleles in the population [23].

Mechanisms Producing FXPOI

Despite abundant literature, all authors agree that only a subset of PM carriers develops FXPOI, however, it remains unknown why some present POI while others do not.

The precise molecular mechanisms of how the *FMR1* premutation causes FXPOI are not well understood, although recent studies have attempted to provide some insight into these mechanisms. Among these reports, a significant positive association of repeat size has been demonstrated with POI and it has been found that women with fewer than 100 repeats have an increased risk of FXPOI, although this relationship is not linear [10, 24]. In this range, it seems that women with 80 to 100 repeats have the highest risk of presenting altered cycle traits and subfertility [20]. Carriers of both smaller and larger PM repeat sizes also present ovarian insufficiency, albeit with a lower frequency. Furthermore, a loss of normal ovarian function has also been demonstrated in women with intermediate alleles [11].

A few studies have described the CGG structure in women with FXPOI, that is, the number of AGG interruptions (See Figure 3 Chapter 1). AGG interruptions have been shown to decrease the risk of the expansion of a PM to a FM, and in one small study it was shown that women with POI and intermediate alleles had no AGG interruptions [13]. Nevertheless, a study by our group [25] did not confirm this observation.

Others studies have postulated the hypothesis that the *FMR1* PM may have a toxic RNA gain-of-function effect on ovarian follicle dynamics. In fact, studies on *FMR1* gene expression in males with the PM showed elevated *FMR1* mRNA levels, and this increased transcriptional activity appeared to be positively correlated with CGG repeat size and with the manifestation of FXTAS [26, 27]. Similarly, the effect of the CGG repeats on the mRNA levels was statistically significant in patients with POF, but not in those without POF

(Figure 3) [10], demonstrating RNA gain-of-function toxicity in PM carrier women.

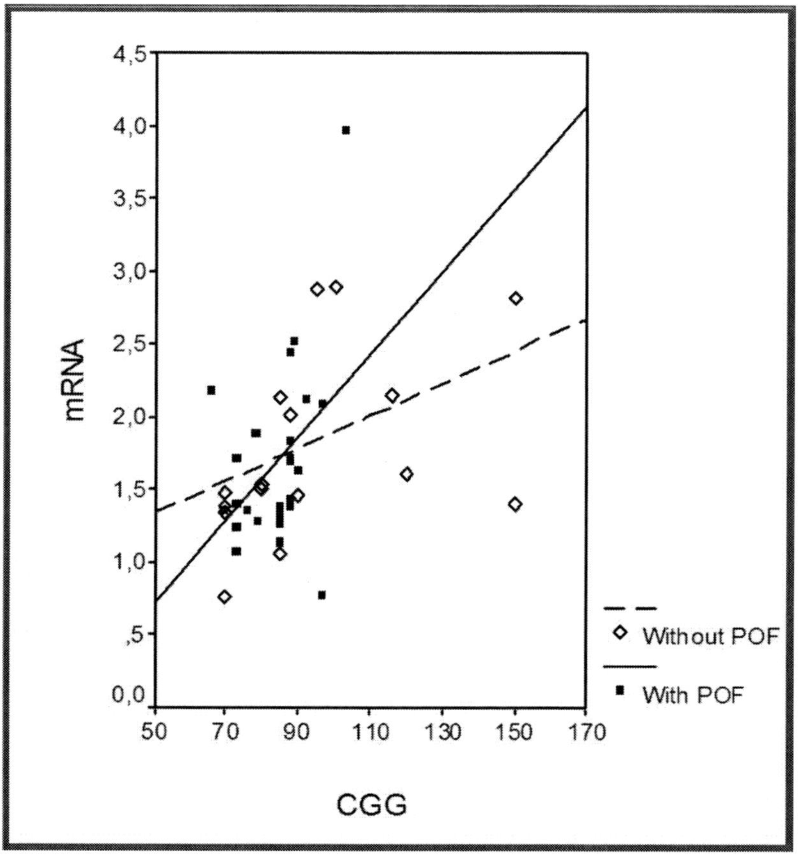

Figure 3. Relation between the CGG repeats and mRNA levels depending on the POF status of the patients. The effect of the CGG repeats on the mRNA levels was statistically significant (p=0.0437) in patients with POF, but not (p=0.0724) in patients without POF (Figure from Tejada et al., 2008 [10]).

However, the relationship between molecular status and phenotype is much more complex in women, in whom the activation ratio of the X chromosome must be considered [9]. Nevertheless, levels of the *FMR1* transcript among female premutation carriers have been demonstrated to be increased, with the relationship between mRNA levels and repeat size being nonlinear, with a cut-off at 100 repeats [10]. It has also been suggested that due to skewed X inactivation mRNA levels tend to normalize in females with an increase in the number of CGG repeats [9].

Indeed, since the *FMR1*gene is located on the X chromosome, it has been proposed that X-chromosome inactivation may play a role in the possible development of FXPOI. However, the first study by Murray et al., [28] reported no significant effect on comparing FXPOI, *FMR1* repeat size and X-chromosome inactivation patterns, having been confirmed by other studies [20].

Regarding FMRP levels, some studies have described increased FMRP levels in the range of 80 to 89 repeats in carrier males [29]; however, similar studies have not been carried out in women [20].

In relation to the parental origin of the PM allele, Hundscheid and co-workers (2000) [30] suggested a role for imprinting on finding a significant effect between FXPOI and parental origin. Nonetheless, further studies have failed to confirm these findings [8].

Finally, FXPOI may not only depend on the PM allele but also on other modifier genes and environmental factors. Hunter et al., [31] showed significant familial aggregation of age at menopause, providing evidence of the presence of additional genes influencing and interacting with FXPOI. The identification of additional genes involved in ovarian function is therefore a new step in future studies. In this regard, Allen and co-workers (2014) [32] performed a study of a candidate gene using whole genome sequencing (WGS) to identify possible variants that may influence the onset of FXPOI. Although no conclusions could be drawn, they did find some potentially damaging variants that deserve further investigation.

In short, to date neither the exact etiology of FXPOI nor the cause of the variation in phenotype is understood, and thus, additional studies are clearly needed.

Mouse Models

Non-invasive methods to study the mechanisms of the PM in ovarian function in humans are currently not available. This is why the development of model systems is necessary to enhance the knowledge of FXPOI. Only recently have mutation vertebrate (mouse and rat) and invertebrate (*Drosophila melanogaster*) models been used to study ovarian function and the results published to date indicate the value of studying the etiology of

FXPOI. The recent work by Sherman and colleagues (2014) [33] reviews and summarizes everything published so far.

Rodent models have basically shown that FMRP is highly expressed in oocytes in which this protein is important for folliculogenesis. The two PM mouse models studied to date [34,35] have showed evidence of ovarian dysfunction and, together, suggest that the long repeat in the transcript itself may have some pathological effect very different from any effect of the toxic protein.

Furthermore, ovarian morphology in young animals appears normal and the primordial follicle pool size does not differ from that of wild-type animals. However, there is a progressive premature decline in the levels of most follicle classes. Observations also include granulosa cell abnormalities and altered gene expression patterns. In this line, in a rat model, Ferder and collaborators (2013) [36] found changes in *FMR1* expression during follicle maturation, both at the protein and mRNA levels. Indeed, in all PM mouse models, increased expression of *FMR1* mRNA has been described in the ovary.

Premutation model systems in non-human primates and those based on induced pluripotent stem cells have shown particular promise and will complement current models, although definite conclusions remain to be established.

Fly models have shown to be very useful because of the relative ease of model construction compared with other model systems. For example, *Drosophila* germ line stem cells have been used as a model to show that FMRP can modulate the fate of stem cells [37]. This study suggested that both the reduction of FMRP and expression of PM CGG repeats could have detrimental effects on fly ovary and stem cell maintenance.

Among other ideas in progress, Sherman and co-workers (2014) [33] have also highlighted the possibility of testing the effect of genetic modifiers on the ovarian phenotype. This could be valuable for not only understanding the pathogenic mechanism, but may also shed light on human homolog genes of which may contribute to the variable penetrance of FXPOI.

In summary, all these studies mentioned above are very promising but definitive conclusions remain to be drawn. Further comparisons of all of these models are still needed to gain insight into the etiology of ovarian dysfunction.

Conclusion

FXPOI is a new clinical entity in which carrier PM females of the *FMR1* gene may present early ovarian dysfunction with menopause occurring at approximately 40 years of age. To date neither the exact etiology of FXPOI nor the cause of the variation in phenotype is understood. New technologies with next generation sequencing and studies in animal models are very promising, but definitive conclusions remain to be drawn, thereby requiring further investigations in this pathology.

References

[1] Hagerman RJ and Hagerman PJ.: Fragile X Syndrome, Diagnosis, treatment and research. Edited by Hagerman and Hagerman. The Johns Hopkins University Press 2002.

[2] Fernández-Carvajal I, Walichiewicz P, et al., Screening for expanded alleles of the FMR1 gene in blood spots from newborns males in a Spanish population. *J. Mol. Diagn.* 2009;11(4):324-9.

[3] Oberlé I, Rousseau F, Heitz D, et al.,: Instability of a 550-base pair DNA segment and abnormal methylation in Fragile X syndrome. *Science.* 1991;252:1097-102.

[4] Rousseau F, Heitz D, Tarleton J, et al.,: A multicenter study on genotype-phenotype correlations in the Fragile X syndrome, using direct diagnosis with probe StB12.3: The first 2.253 cases. *Am. J. Hum. Genet.* 1994;55:225-237.

[5] Nolin SL, Brown WT, Glicksman A, et al.,: Expansion of the Fragile X CGG repeat in females with Premutation or Intermediate alleles. *Am. J. Hum. Genet.* 2003;72:454-464.

[6] Vianna-Morgante AM, Costa SS, Pares AS, et al.,: FRAXA premutation associated with premature ovarian failure. *Am. J. Med. Genet.* 1996;64:373-375.

[7] Allingham-Hawkins DJ, Babul-Hirji R, Chitayat D, et al., Fragile X premutation is a significant risk factor for premature ovarian failure: the International Collaborative POF in Fragile X study-preliminary data. *Am. J. Med. Genet.* 1999;83:322-325.

[8] Mallolas J, Duran M, Sanchez A, et al.,: Implications of the FMR1 gene in menopause: study of 147 Spanish women. *Menopause.* 2001;8(2):106-10.

[9] García-Alegría E, Ibáñez B, Mínguez M, et al., Analysis of FMR1 gene expression in female premutation carriers using robust segmented linear regression models. *RNA.* 2007;13:756-62.

[10] Tejada MI, García-Alegría E, Bilbao A, et al.,: Analysis of the molecular parameters that could predict the risk of manifesting POF in female PM carriers of fragile X syndrome. *Menopause.* 2008;15(5):945-9.

[11] Wittenberger MD, Hagerman RJ, Sherman SL, et al., The FMR1 premutation and reproduction. *FertilSteril.* 2007;87:456-65.

[12] Murray A, Schoemaker MJ, Bennet CE, et al., Population-based estimates of the prevalence of FMR1expansion mutations in women with early menopause and primary ovarian insufficiency. *Genetics in Medicine.* 2014;16:19-24.

[13] Bodega B, Bione S, Dalpra L, et al.,: Influence of intermediate and uninterrupted FMR1 CGG expansions in premature ovarian failure manifestation. *Hum. Reprod.* 2006;21:952–7.

[14] Gleicher N, Weghofer A and Barad DH.: A pilot study of premature ovarian senescence: I. Correlation of triple CGG repeats on the FMR1 gene to ovarian reserve parameters FSH and anti-Müllerian hormone. *Fertil Steril.* 2009;91:1700-1706.

[15] Nelson LM. Menstrual cycles, the FMR1 Premutation an "Primary Ovarian Insufficiency". *The National Fragile X Foundation Quarterly.* 2007;26:22-23.

[16] De Caro JJ, Dominguez C and Sherman SL.: Reproductive health of adolescent girls who carry the FMR1 Premutation: Expected phenotype based on current knowledge of Fragile X-Associated primary ovarian insufficiency. *Ann. NY Acad. Sci.* 2008;1135:99-111.

[17] Nelson LM, Covington SN, and Rebar RW.: An update: spontaneous premature ovarian failure is not an early menopause. *Fertil Steril* 2005;83:1327-1332.

[18] Nelson LM. Clinical practice. Primary ovarian insufficiency. *N. Engl. J. Med.* 2009;360:606-614.

[19] Welt CK, Smith PC and Taylor AE.: Evidence of early ovarian aging in Fragile X premutation carriers. *J. Clin. Endocrinol. Metab.* 2004;89:4569-74.

[20] Sullivan SD, Welt C and Sherman S.: FMR1 and the Continuum of Primary Ovarian Insufficiency. *Semin. Reprod. Med.* 2011;29:299-307.

[21] Tejada MI.: Manifestaciones clínicas en las mujeres con premutación. Fallo Ovárico Prematuro (POF). En: Síndrome X Frágil. Libro de consulta para familias y profesionales, pags. 39-44. *Edita: Real Patronato sobre discapacidad,* Madrid 2006.

[22] Santoro N. Research on the mechanisms of premature ovarian failure. *J. Soc. Gynecol. Investig.* 2001;8:S10-2.

[23] Crawford DC, Acuña JM, Sherman SL. FMR1 and the fragile X syndrome: human genome epidemiology review. *Genet Med.* 2001;3:359-371.

[24] Ennis S, Ward D, Murray A. Nonlinear association between CGG repeat number and age of menopause in FMR1 premutation carriers. *Eur. J. Hum. Genet.* 2006;14:253-5.

[25] Tejada MI, Maortua H, Mendoza R, et al., Study of triple CGG repeats (FMR1 gene) in 41 patients with a reduced ovarian response to gonadotropin stimulation: no correlation found with four ovarian parameters. *Hum. Reprod.* 2013;28:241-2.

[26] Tassone F, Hagerman RJ, Taylor AK, et al.,.Elevated Levels of FMR1 mRNA in Carrier Males: A New Mechanism of involvement in the Fragile-X Syndrome. *Am. J. Hum. Genet.* 2000;66:6–15.

[27] Jacquemont S, Hagerman RJ, Leehey M, et al., Fragile X premutation tremor/ataxia syndrome: molecular, clinical, and neuroimaging correlates. *Am. J. Hum. Genet.* 2003;72:869-78.

[28] Murray A, Ennis S, Mac Swiney F, et al., Reproductive and menstrual history of females with fragile X expansions. *Eur. J. Hum. Genet.* 2000;8:247-52.

[29] Peprah E, He W, Allen E et al., Examination of FMR1 transcript and protein levels among 74 premutation carriers. *J. Hum. Genet.* 2010;55:66-68.

[30] Hundscheid RD, Sistermans EA, Thomas CM, et al., Imprinting effect in premature ovarian failure confined to paternally inherited fragile X premutations. *Am. J. Hum. Genet.* 2000;66:413-28.

[31] Hunter JE, Epstein MP, Tinker SW, et al., Fragile X-associated primary ovarian insufficiency: evidence for additional genetic contributions to severity. *Genet Epidemiol.* 2008;32:553-559.

[32] Allen EG, Grus WE, Narayan S, et al., Approaches to identify genetic variants that influence the risk for onset of fragile X-associated primary ovarian insufficiency (FXPOI): a preliminary study. *Frontiers in Genetics.* 2014;5:1-8.

[33] Sherman S, Curnow EC, Easley ChA, et al., Use of model systems to understand the etiology of fragile X-associated primary ovarian insufficiency (FXPOI). *Journal of Neurodevelopmental Disorders*. 2014; 6:26.

[34] Lu C, Lin L, Tan H, et al., Fragile X premutation RNA is sufficient to cause primary ovarian insufficiency in mice. *Hum. Mol. Genet.* 2012, 21:5039–5047.

[35] Ascano M Jr, Mukherjee N, Bandaru P, et al., FMRP targets distinct mRNA sequence elements to regulate protein expression. *Nature*. 2012, 492:382–386.

[36] Ferder I, Parborell F, Sundblad V, et al., Expression of fragile X mental retardation protein and FMR1 mRNA during folliculogenesis in the rat. *Reprod Abstr. Ser*. 2013, 145:335–343.

[37] Yang L, Duan R, Chen D, et al., Fragile X mental retardation protein modulates the fate of germline stem cells in Drosophila. *Hum. Mol. Genet.* 2007, 16:1814–20.

In: Allelic Forms of the FMR1 Gene
Editor: Montserrat Milà

ISBN: 978-1-63321-914-4

Chapter 3

Fragile X-Associated Tremor/Ataxia Syndrome

L. Rodriguez-Revenga
Biochemistry and Molecular Genetics
Department. Hospital Clínic. Barcelona, Spain
CIBER de Enfermedades Raras (CIBERER), Barcelona, Spain
IDIBAPS (Institut d'Investigacions Biomèdiques August Pi i Sunyer).
Barcelona, Spain

Abstract

Fragile X-associated tremor/ataxia syndrome (FXTAS) is a late-onset inherited neuropsychiatric degenerative disorder that occurs predominantly in male carriers of the *FMR1* premutation (55-200 CGG repeats). *FMR1* premutation is relatively frequent in the general population, affecting approximately 1 out of 800 males and 1 out of 250 females, and leading to symptoms of FXTAS in up to 1 in 3,000 men older than 50 years. Clinical symptoms in FXTAS patients usually begin with an action tremor. After that, different findings including ataxia (balance problems with frequent falling), and more variably, loss of sensation in the distal lower extremities and autonomic dysfunction (e.g., impotence, hypertension, and loss of bowel and bladder function), may occur, and gradually progress. Molecular mechanism leading to FXTAS is distinct from the *FMR1* silencing mechanism and/or a deficit in FMRP operating in fragile X syndrome. Individuals with *FMR1* premutation alleles have markedly elevated levels of expanded CGG-repeat *FMR1* mRNA, which is thought to have a toxic gain-of-function. Since 2001, when FXTAS was first described, the advancement of our understanding of the clinical phenotype as well as the molecular pathophysiology has occurred very quickly. The aim of this chapter is to present the most recent advances in the current knowledge of FXTAS.

Keywords: *FMR1*, FXTAS, ataxia, tremor, mRNA toxicity

Introduction

Fragile X-associated tremor/ataxia syndrome, FXTAS (OMIM# 300623) was identified in 2001 by Hagerman and co-workers as a late-onset neurodegenerative disorder [1]. It took more than 10 years after the *FMR1* gene identification to recognize FXTAS as a *FMR1* permutation associated phenotype. One explanation for this is that the movement disorder experienced by older carriers, who were thought to be clinically normal, was not associated with the fragile X syndrome (FXS) (a childhood disorder) affecting children. Mothers of FXS children, being seen in clinics, were often expressing concerns about their fathers (*FMR1* premutation carriers) who were experiencing problems with hand tremor and unsteady gait. When Hagerman and co-workers evaluated these male carriers they found that they were more likely to develop, from 50 years onwards, a stereotyped clinical picture characterized by unsteadiness while walking and action tremor in both hands. Both symptoms used to follow a progressive course, and were often accompanied by progressive cognitive and behavioral disturbances [1]. Although, FXTAS was firstly identified among older *FMR1* premutation male carriers, to date it is well established that it also affects females. Nevertheless, it has been suggested that it occurs less frequently and that the phenotype is milder with older age at onset [1, 2]. An explanation for this difference is the presence of a second normal allele and a random X-inactivation of the premutated one; however, there may be additional sex-specific effects that reduce penetrance among females [3].

FXTAS penetrance is not complete; meaning that not all *FMR1* premutation carriers develop FXTAS. It is largely dependent on carrier age, with symptoms typically seen in 30% of males by the age of 50 and in 50% of males older than 70 years. Moreover, there is significant variability in the progression of neurological dysfunction [2-5].

The description and characterization of FXTAS syndrome is of great interest to the population, because the prevalence of *FMR1* premutation in the general population is relatively high. In fact, it is estimated that 1 in 800 to 1,200 males and from 1 in 250 to 400 females are carriers of a *FMR1* premutation allele [6-8]. Although the contribution of FXTAS to the morbidity and mortality of the aging population requires further study, the disorder is likely the most common single-gene form of tremor and ataxia in the older adult population. Taking into consideration the prevalence of *FMR1* premutation in the general population, the prevalence of FXTAS might be estimated in ~1/3,000 males aged over 50 years of age (~1/10,000 males of all ages) [3].

Fragile X-Associated Tremor/Ataxia Syndrome

FXTAS Clinical and Cognitive Overview

Originally FXTAS was clinically described as a motor disorder with core features including kinetic tremor and cerebellar gait ataxia. Clinical symptoms of FXTAS syndrome appear in patients in their 50s and they usually begin with an action tremor. Additional clinical features include peripheral neuropathy (60%), impotence (80%), bowel and bladder dysfunction (30-55%), erectile dysfunction, loss of sensation in the distal lower extremities, hearing loss and dysphagia [reviewed in 9].

Over the last years, the clinical spectrum of the disease has expanded and new data have evidenced that progressive cognitive decline, with a gradual progression to dementia in some individuals, and neurophsychological problems are common among FXTAS patients [10-16]. Several studies comparing psychiatric symptoms in FXTAS with matched comparison groups have been performed. Overall, they describe that the neuropsychiatric phenotype of FXTAS is characterized primarily by poor performance on measures of executive function, working memory, information processing speed, and fine motor control. While attention and language function may also be affected in patients with FXTAS, there is limited evidence regarding the affectation of visuospatial function. Secondary behavioral features may include depression, anxiety, and obsessive-compulsive symptoms. Finally all evidences concluded that this profile is consistent with a dysexecutive fronto-subcortical syndrome [10, 16, 17].

FXTAS clinical symptoms tends to progress from mild ataxia and/or tremor to a disabling condition impacting on motor daily activities, thinking, and social skills that seriously compromise patients' quality of life.

Because of the variety of clinical symptoms in FXTAS, genetic testing for FXTAS must be considered in any patient (male and female) with ataxia developing after the age of 50. Diagnostic testing consists of determining *FMR1* CGG repeat size by polymerase chain reaction (PCR) technique, which nowadays is a highly sensitive test performed in many laboratories.

FXTAS Neuroimaging and Neuropathological Profiles

Magnetic resonance imaging (MRI) of the brain was obtained in FXTAS patients searching for specific features. Individuals with FXTAS demonstrate moderate to severe generalized brain atrophy with ventricular enlargement, cerebellar atrophy, and subcortical and/or pontocerebellar white matter lesions [18, 19]. This pattern frequently includes hyperintensities in the cerebellar white matter and middle cerebellar peduncles on T2-weighted images, called the "MCP sign" [19]. These characteristic findings, initially described on conventional MRI, were classified into two categories -major and minor criteria-, and proposed, together with clinical findings, as diagnostic criteria for FXTAS (Table 1) [18].

Cerebellar abnormalities were confirmed by post-mortem histological studies that identified several neuropathological features, including Purkinje cell decreases and spongiform changes [20, 21]. Significant loss in whole cerebellar volume has been revealed in both male and female patients with FXTAS [22, 23].

Although the MCP sign is considered a major radiological feature of FXTAS, it has been also reported in patients with other forms of adult-onset cerebellar ataxia, thus resting specificity for FXTAS [24]. As more evidence among FXTAS individuals developing neurological features is being gained, the spectrum and variability of MRI features becomes broader and the potential usefulness of neuroimaging in providing insight in FXTAS becomes more evident [reviewed in 25]. Furthermore, with the advent of high field strength MRI as well as more sophisticated postprocessing tools the understanding of FXTAS neuropathological profile becomes wider. In this context, Hashimoto and co-workers [14] assessed focal changes of grey and white matter density in the brain of *FMR1* premutated carriers using voxel based morphometry method. In this study, patients with FXTAS demonstrated a distinct pattern of grey matter volume loss, involving multiple cortical and subcortical regions. This included different parts of the cerebellum, as well as of the medial surface of the brain, including the dorsomedial prefrontal cortex, anterior cingulate and precuneus. Additional volume loss was seen in the lateral prefrontal cortex, orbitofrontal cortex, amygdala, and insula. More interestingly, there were significant correlations between grey matter loss in different brain regions, behavioral scales, and CGG repeats [14].

Table 1. Clinical criteria for FXTAS

Examination and Degree	Observation
Radiological:	
Major	MRI white matter lesions in MCPs and or brain stem
Minor	MRI white matter lesions in cerebral white matter
Minor	Moderate-to-severe generalized atrophy
Clinical:	
Major	Intention tremor
Major	Gait ataxia
Minor	Parkinsonism
Minor	Moderate-to-severe short-term memory deficiency
Minor	Executive function deficit

Inclusion criterion: CGG repeat number between 55 and 200.
Note. Data described by [18].

These radiologic features have been histologically confirmed in FXTAS post mortem cases. Early neuropathological, post mortem studies of the brain of patients with FXTAS revealed intranuclear inclusions in neurons and astrocytes throughout the cerebrum and brainstem, being most numerous in the hippocampal formation [20]. These inclusions appear as eosinophilic, hyaline, refractile, 2-5 μm diameter, round to ovoid bodies that show positive reactivity with antibodies against over 20 different proteins including ubiquitin, αβ-crystallin, lamin A/C, hnRNP A2, myelin basic protein, DNA repair-ubiquitin-associated HR23B, and Sam68, among others [26-28]. The inclusions are PAS, silver, amyloid, and α-synuclein negative [20, 21]. Additionally, the

FMR1 mRNA but not the FMRP has been found contained within intranuclear inclusions [27, 29].

Although Purkinje cells rarely have inclusions, there is neurodegeneration in the cerebellum, with marked Purkinje cell loss, axonal swelling, and gliosis. Apart from reduced Purkinje cell number, additional neuropathological features present in FXTAS include, axonal torpedoes, and prominent cortical and subcortical white matter pathology [20, 21].

Recently, a broad distribution of intranuclear inclusions in non-central nervous system has been observed in *FMR1* premutation carriers with FXTAS [28]. Inclusions were found in somatic organs such as the endocrine organs, gastrointestinal tract, heart, and kidney [28]. Although these findings are consistent with the expanding range of co-morbid medical features reported in FXTAS (see chapter 5), intranuclear inclusions are not the cause of these conditions.

Nowadays FXTAS is considered as a new class of inclusion disorder. However there are several unclear aspects such as the cellular processes underling inclusion formation and whether the inclusions are themselves toxic or simply reflect underlying cellular dysfunction.

Molecular Genetics Overview

FXTAS is an allelic disorder to the FXS, and therefore should be considered as a distinct neurodegenerative disorder. Although both disorders are caused by expansions of the CGG repeat element found in the *FMR1* gene, the molecular mechanism leading to FXTAS is distinct from the *FMR1* silencing mechanism and/or a deficit in FMRP operating in FXS. In premutated patients the *FMR1* gene is rarely silenced and FMRP levels are generally normal or only slightly lowered [3]. The only known molecular abnormality among *FMR1* premutation carriers is the presence of markedly elevated levels (~2-8 fold) of *FMR1* mRNA.

The increased transcriptional activity of the *FMR1* gene seems to be positively correlated with the size of the CGG repeat. That is, CGG repeats in the upper range (100-200 CGG) result in average 5-8 fold elevation, whereas CGGs in the lower range (50-100 CGG) result in an average 2-fold elevation [30-33]. Although the precise mechanism for this overexpression is unknown, several possible mechanisms have been postulated. A feedback mechanism suggests that the cell attempts to compensate for reduced levels of FMRP by increasing the amount of available *FMR1* transcript [reviewed in 34, 35].

Alternatively, it is likely that the increasing length of the CGG repeat near the *FMR1* promoter proportionally opens the chromatin, allowing more ready access to transcription factors [36]. The presence of these elevated levels of abnormal (expanded CGG repeat) *FMR1* mRNA led to propose an RNA "toxic gain-of-function" model for FXTAS, in which the mRNA itself, with the abnormal CGG repeat tract, is causative of the neurological disorder [1, 3, 18, 20]. Animal and cell-based studies have demonstrated that the sole expression of expanded CGG repeats is necessary and sufficient to cause pathology similar to human FXTAS [37-40], and thus indicate that the expanded CGG repeats in RNA are the likely cause of the neurodegeneration in FXTAS.

In particular, the hypothesis of a RNA toxic gain-of-function suggests that CGG repeat *FMR1* mRNA recruits RNA-binding proteins and other proteins through direct RNA–protein interactions, but probably also through indirect protein–protein interactions. The sequestration of these proteins is also thought to prevent normal function leading to downstream alterations [34, 41].

Nowadays, several proteins have been identified that interact with the CGG repeat and might be potential mediators of downstream cellular dysregulation [reviewed in 42]. Table 2 summarizes candidate protein mediators of the CGG-repeat mRNA toxicity. The protein components of FXTAS inclusions fell into eight major functional categories, including: histone family; intermediate filament; microtubule; myelin-associated proteins; RNA-binding proteins; stress-related proteins; chaperones and ubiquitin-proteasome-related proteins [reviewed in 34]. Although most of them have been found to localize with ubiquitin-positive inclusions in CGG-expressing Drosophila, knock-in (KI) mouse model and FXTAS patients, they are not found to be sequestered by expanded CGG repeats and consequently they are not expected to lose their functions in FXTAS patients [43]. The lack of a principal protein species in the inclusions is not surprising, since there is no known abnormal protein product associated with FXTAS. In particular, the protein product of the *FMR1* gene, FMRP, is not structurally abnormal, as the expanded CGG repeat is located in a non-coding portion (5′-UTR, 5′-untranslated region) of the gene. A striking characteristic of CGG expanded repeats is that they form dynamic intranuclear RNA aggregates that enlarge with time, resulting in the formation of giant inclusions. Continuous enlargement of CGG RNA aggregates suggests that these repeats may constantly recruit proteins, implying a founding RNA-protein interaction event that would subsequently trap other proteins through indirect RNA-protein or protein-protein interactions [43].

Table 2. Candidate proteins involved in mediating the RNA-triggered pathogenesis

Protein	Function	Evidences	Reference
Sam68	Splicing modulator	- its ablation in mouse leads to motor dysfunction - is partially sequestered and functionally inactivated in FXTAS patients	[43, 44]
Pur α	Single-stranded cytoplasmatic DNA and RNA binding protein	- mediates neurodegeneration in Drosophila and mouse - binds CGG repeats in both mammalian and Drosophila brains - found to be present in the inclusions of FXTAS brain tissues - overexpression can alleviate neurodegeneration in FXTAS fly model	[45]
hnRNP A1 hnRNP A2/B1	RNA binding proteins	- assists mRNA dendritic transport in drosophila and cultured rat neurons - binds CGG repeats in both mammalian and Drosophila brains - found to be present in the inclusions of human FXTAS brain tissues - overexpression suppresses the phenotype of the CGG transgenic fly	[27, 46, 47]
DROSHA-DGCR8	microRNA nuclear processing complex	- binds to expanded CGG repeats - microRNA dysregulation in human FXTAS brain tissue - overexpression of DGCR8 rescues neuronal cell death	[48]
ubiquitin	post-translational modification	- present in the inclusions of human FXTAS brain samples	[27]
αB-crystallin	heat shock protein	- present in the inclusions of human FXTAS brain samples	[27]
Lamin A/C	intermediate filament proteins	- present in the inclusions of human FXTAS brain samples	[27]
MBP	myelin basic protein	- present in the inclusions of human FXTAS brain samples	[27]

Proteins recruited through protein-protein interactions have not been listed (e.g. CUGBP1, Rm62, Hsp70).

The current hypothesis contemplates that the CGG-repeat expansion drive to cellular dysregulation that cause eventual cell death. Mitochondrial abnormalities [49, 50], altered calcium regulation [51], altered lamin A nuclear architecture [52], and reduced telomere length [53], have been described as potential downstream pathways that could mediate cellular dysregulation [reviewed in 54].

Although FXATS is defined as a late-onset disorder, recent findings prompt to a process that begins in early development. Abnormalities of neuronal morphology and network activity as well as alerted calcium regulation have found in late embryonic and neonatal premutation mouse model [51, 55-57]. This observation suggests that there is an early developmental impairment that precedes cellular loss of viability, which somehow might explain the broad clinical phenotype associated with the *FMR1* premutation (See chapter 5).

As exposed in this chapter, there are several evidences that support the RNA-toxic-gain of function model for FXTAS. However, at least two additional mechanisms of toxicity have been described [reviewed in 58]. The first one compromises the translation in all reading frames of expanded CGG repeats in the absence of an ATG initiation codon (non-AUG initiated translation, RAN), leading to the production of homopolymeric protein. In particular, Todd and co-workers [59] demonstrated the existence of one particular CGG RAN translation product (polyglycine), which directly modulates CGG-associated pathology in two distinct model systems. Therefore on the basis of their observations it is suggested that a protein gain-of-function may also occur in cells of patients with FXTAS. Secondly, an antisense *FMR1* mRNA and a number of long noncoding RNAs have been identified within/near the *FMR1* gene [60-62]. Although the relevance of these transcripts is not fully understood they might have important functions and/or modulate certain aspects of FXTAS [reviewed in 58].

Overall, these observations raise the possibility of different or additive/synergistic molecular mechanisms contributing to the pathogenesis of FXTAS.

Even though the understanding of molecular basis of FXTAS has evolved over the last years and many different aspects of its pathogenesis have been unraveled, we still do not know how to link all of them. Furthermore, there is the puzzling issue of the incomplete penetrance. Additional genetic and/or environmental factors might play a key role and might be related with the emergence of FXTAS clinical symptoms. In this scenario, further research is needed since the better understanding off all these aspects might help to prevent or delay the onset of the syndrome.

FXTAS Treatment

Currently, there is no specific targeted treatment for patients with FXTAS. The present medical management of patients with FXTAS is limited to medications that targets specific symptoms such as tremor, ataxia, mood changes, depression, cognitive decline and/or dementia [reviewed in 9, 54]. Although neurological rehabilitative therapies have not been studied specifically in FXTAS they should be considered in treatment. Furthermore, occupational and physical therapy might also be beneficial and improve strength and stability in those with tremor, gait and balance deficits.

RNA toxicity initiates brain changes well before the onset of FXTAS. Typically, the age of onset of FXTAS is in the early 60s, and it presents with onset of tremor. However, as described before in this chapter different findings suggest that asymptomatic changes associated with the pathological changes of FXTAS exist before the onset of tremor or ataxia. Additional genetic and/or environmental factors might be necessary to precipitate FXTAS symptoms. In this regard, chemotherapy, prolonged surgery with general anesthesia or exposure to toxins, genetic predisposition to mitochondrial dysfunction, coexistence with another disease such as multiple sclerosis or Alzheimer's disease have been described as factors that can precipitate FXTAS clinical manifestations [reviewed in 54].

On the basis of these observations, FXTAS treatment should envision two approaches; a direct one addressed to eliminate the expanded CGG-repeat mRNA and a prophylactic one, protecting neurological cells from neurotoxic chemicals.

Conclusion

Since 2001, when FXTAS was first described, the advancement of our understanding of the clinical phenotype as well as the molecular pathophysiology has occurred very quickly. Nowadays, FXTAS clinical description includes a much broader symptoms spectrum than the initial core features of intention tremor and gait ataxia. Furthermore, since the observation of increased expression of *FMR1* mRNA, a RNA toxicity hypothesis and other possible pathological mechanisms have been proposed with multiple studies supporting them. Although a lot has been achieved in this short period of time, further work is necessary to link and fully understand all the molecular mechanisms that might play a role in FXTAS pathology. The better understanding of these processes should also shed light on therapeutic approaches that will combat not only neurodegeneration but also the rest of symptoms. Finally, as the prevalence of *FMR1* premutated alleles is relatively high in general population, FXTAS may represent one of the more common monogenic causes of tremor, ataxia, and dementia. For this reason, it is probably that many carriers with FXTAS are being seen by a clinical specialist without awareness of the underlying genetic basis for the symptoms. The early diagnosis of those patients not only benefits themselves but also the rest of the family that should be advised for the FXS.

References

[1] Hagerman RJ, Leehey M, Heinrichs W, et al. Intention tremor, parkinsonism, and generalized brain atrophy in male carriers of fragile X. *Neurology*. 2001; 57: 127-130.

[2] Jacquemont S, Farzin F, Hall D, et al. Aging in individuals with the FMR1 mutation. *Am. J. Ment. Retard*. 2004; 109: 154-164.

[3] Hagerman PJ, Hagerman RJ. The fragile-X premutation: a maturing perspective. *Am. J. Hum. Genet*. 2004; 74: 805-816.

[4] Jacquemont S, Hagerman RJ, Leehey MA, et al. Penetrance of the fragile X-associated tremor/ataxia syndrome in a premutation carrier population. *JAMA*. 2004; 291: 460-469.

[5] Rodriguez-Revenga L, Madrigal I, Pagonabarraga J, et al. Penetrance of FMR1 premutation associated pathologies in fragile X syndrome families. *Eur. J. Hum. Genet*. 2009; 17: 1359-1362.

[6] Rousseau F, Rouillard P, Morel ML et al. Prevalence of carriers of premutation-size alleles of the FMRI gene--and implications for the population genetics of the fragile X syndrome. *Am. J. Hum. Genet*. 1995; 57: 1006-1018.

[7] Dombrowski C, Lévesque S, Morel ML, et al. Premutation and intermediate-size FMR1 alleles in 10572 males from the general population: loss of an AGG interruption is a late event in the generation of fragile X syndrome alleles. *Hum. Mol. Genet*. 2002; 11: 371-378.

[8] Rifé M, Badenas C, Mallolas J, et al. Incidence of fragile X in 5,000 consecutive newborn males. *Genet Test.* 2003; 7: 339-343.

[9] Hall DA, O'Keefe JA. Clinical neurogenetics: fragile x-associated tremor/ataxia syndrome. *Neurol. Clin.* 2013; 31: 1073-1084.

[10] Bacalman S, Farzin F, Bourgeois JA, et al. Psychiatric phenotype of the fragile X-associated tremor/ataxia syndrome (FXTAS) in males: newly described fronto-subcortical dementia. *J. Clin. Psychiatry.* 2006; 67: 87-94.

[11] Bourgeois JA, Cogswell JB, Hessl D, et al. Cognitive, anxiety and mood disorders in the fragile X-associated tremor/ataxia syndrome. *Gen. Hosp. Psychiatry.* 2007; 29: 349-56.

[12] Bourgeois JA, Seritan AL, Casillas EM, et al. Lifetime prevalence of mood and anxiety disorders in fragile X premutation carriers. *J. Clin. Psychiatry.* 2011; 72: 175-82.

[13] Adams PE, Adams JS, Nguyen DV, et al. Psychological symptoms correlate with reduced hippocampal volume in fragile X premutation carriers. *Am. J. Med. Genet. B Neuropsychiatr Genet.* 2010; 153B: 775-785.

[14] Hashimoto R1, Backer KC, Tassone F, et al. An fMRI study of the prefrontal activity during the performance of a working memory task in premutation carriers of the fragile X mental retardation 1 gene with and without fragile X-associated tremor/ataxia syndrome (FXTAS). *J. Psychiatr. Res.* 2011; 45: 36-43.

[15] Hessl D, Tassone F, Loesch DZ, et al. Abnormal elevation of FMR1 mRNA is associated with psychological symptoms in individuals with the fragile X premutation. *Am. J. Med. Genet. B Neuropsychiatr Genet.* 2005; 139B: 115-121.

[16] Birch RC, Cornish KM, Hocking DR, Trollor JN. Understanding the Neuropsychiatric Phenotype of Fragile X-Associated Tremor Ataxia Syndrome: a Systematic Review. *Neuropsychol .Rev.* 2014.

[17] Brega AG, Goodrich G, Bennett RE, et al. The primary cognitive deficit among males with fragile X-associated tremor/ataxia syndrome (FXTAS) is a dysexecutive syndrome. *Clin. Exp. Neuropsychol.* 2008; 30: 853-869.

[18] Jacquemont S, Hagerman RJ, Leehey M, et al. Fragile X premutation tremor/ataxia syndrome: molecular, clinical, and neuroimaging correlates. *Am. J. Hum. Genet.* 2003; 72: 869-878.

[19] Brunberg JA, Jacquemont S, Hagerman RJ, et al. Fragile X premutation carriers: characteristic MR imaging findings of adult male patients with

progressive cerebellar and cognitive dysfunction. *AJNR Am. J. Neuroradiol.* 2002; 23: 1757-1766.

[20] Greco CM, Hagerman RJ, Tassone F, et al. Neuronal intranuclear inclusions in a new cerebellar tremor/ataxia syndrome among fragile X carriers. *Brain*. 2002; 125: 1760-1771.

[21] Greco CM, Berman RF, Martin RM, et al. Neuropathology of fragile X-associated tremor/ataxia syndrome (FXTAS). *Brain*. 2006; 129: 243-255.

[22] Cohen S, Masyn K, Adams J, et al. Molecular and imaging correlates of the fragile X-associated tremor/ataxia syndrome. *Neurology*. 2006; 67: 1426-1431.

[23] Adams JS, Adams PE, Nguyen D, et al. Volumetric brain changes in females with fragile X-associated tremor/ataxia syndrome (FXTAS). *Neurology*. 2007; 69: 851-859.

[24] Okamoto K, Tokiguchi S, Furusawa T, et al. MR features of diseases involving bilateral middle cerebellar peduncles. *AJNR Am. J. Neuroradiol.* 2003; 24: 1946-1954.

[25] Rodriguez-Revenga L, Gómez-Ansón B, Granell Moreno E, et al. Neuroimaging in Fragile X-Associated, Neuroimaging-Clinical Applications, Prof. Peter Bright (Ed.), 2012.

[26] Bergink S, Severijnen LA, Wijgers N, et al. The DNA repair-ubiquitin-associated HR23 proteins are constituents of neuronal inclusions in specific neurodegenerative disorders without hampering DNA repair. *Neurobiol. Dis.* 2006; 23: 708-716.

[27] Iwahashi CK, Yasui DH, An HJ, et al. Protein composition of the intranuclear inclusions of FXTAS. *Brain*. 2006; 129: 256-271.

[28] Hunsaker MR, Greco CM, Spath MA, et al. Widespread non-central nervous system organ pathology in fragile X premutation carriers with fragile X-associated tremor/ataxia syndrome and CGG knock-in mice. *Acta Neuropathol.* 2011; 122: 467-479.

[29] Tassone F, Iwahashi C, Hagerman PJ. FMR1 RNA within the intranuclear inclusions of fragile X-associated tremor/ataxia syndrome (FXTAS). *RNA Biol.* 2004; 1: 103-105.

[30] Kenneson A, Zhang F, Hagedorn CH, Warren ST. Reduced FMRP and increased FMR1 transcription is proportionally associated with CGG repeat number in intermediate-length and premutation carriers. *Hum. Mol. Genet.* 2001; 10: 1449-1454.

[31] Oostra BA, Willemsen R. A fragile balance: FMR1 expression levels. *Hum. Mol. Genet.* 2003; 12: R249-R257.

[32] Tassone F, Hagerman RJ, Taylor AK, et al. Elevated levels of FMR1 mRNA in carrier males: a new mechanism of involvement in the fragile-X syndrome. *Am. J. Hum. Genet.* 2000; 66: 6-15.

[33] Tassone F, Hagerman RJ, Taylor AK, et al. Clinical involvement and protein expression in individuals with the FMR1 premutation. *Am. J. Med Genet.* 2000; 91: 144-152.

[34] Galloway JN, Nelson DL. Evidence for RNA-mediated toxicity in the fragile X-associated tremor/ataxia syndrome. *Future Neurol.* 2009; 4: 785.

[35] Tassone F, Hagerman PJ. Expression of the FMR1 gene. *Cytogenet Genome Res.* 2003; 100: 124-128.

[36] Tan H, Li H, Jin P. RNA-mediated pathogenesis in fragile X-associated disorders. *Neurosci. Lett.* 2009; 466: 103-108.

[37] Jin P, Zarnescu DC, Zhang F, et al. RNA-mediated neurodegeneration caused by the fragile X premutation rCGG repeats in Drosophila. *Neuron.* 2003; 39: 739-747.

[38] Arocena DG, Iwahashi CK, Won N, et al. Induction of inclusion formation and disruption of lamin A/C structure by premutation CGG-repeat RNA in human cultured neural cells. *Hum. Mol. Genet.* 2005; 14: 3661-3671.

[39] Hashem V, Galloway JN, Mori M, et al. Ectopic expression of CGG containing mRNA is neurotoxic in mammals. *Hum. Mol. Genet.* 2009; 18: 2443-2451.

[40] Hukema RK, Buijsen RAM, RaskeC et al. Induced expression of expanded CGG RNA causes mitochondrial dysfunction in vivo. *Cell Cycle.* 2014: 16: 2600-2608.

[41] Li Y, Jin P. RNA-mediated neurodegeneration in fragile X-associated tremor/ataxia syndrome. *Brain Res.* 2012; 1462: 112-117.

[42] Hagerman PJ. Current Gaps in Understanding the Molecular Basis of FXTAS. *Tremor Other Hyperkinet Mov* (N Y). 2012; 2.

[43] Sellier C, Rau F, Liu Y, et al. Sam68 sequestration and partial loss of function are associated with splicing alterations in FXTAS patients. *EMBO J.* 2010; 29: 1248-1261.

[44] Lukong KE, Richard S. Motor coordination defects in mice deficient for the Sam68 RNA-binding protein. *Behav. Brain Res.* 2008; 189: 357-363.

[45] Jin P, Duan R, Qurashi A, et al. Pur alpha binds to rCGG repeats and modulates repeat-mediated neurodegeneration in a Drosophila model of fragile X tremor/ataxia syndrome. *Neuron.* 2007; 55: 556-564.

[46] Sofola OA, Jin P, Qin Y et al. RNA-binding proteins hnRNP A2/B1 and CUGBP1 suppress fragile X CGG premutation repeat-induced neurodegeneration in a Drosophila model of FXTAS. *Neuron*. 2007; 55: 565-571.

[47] Muslimov IA, Patel MV, Rose A, Tiedge H. Spatial code recognition in neuronal RNA targeting: role of RNA-hnRNP A2 interactions. *J. Cell Biol.* 2011; 194: 441-457.

[48] Sellier C, Freyermuth F, Tabet R, et al. Sequestration of DROSHA and DGCR8 by expanded CGG RNA repeats alters microRNA processing in fragile X-associated tremor/ataxia syndrome. *Cell Rep*. 2013; 3: 869-880.

[49] Napoli E, Ross-Inta C, Wong S, et al. Altered zinc transport disrupts mitochondrial protein processing/import in fragile X-associated tremor/ataxia syndrome. *Hum. Mol. Genet*. 2011; 20: 3079-3092.

[50] Ross-Inta C, Omanska-Klusek A, Wong S, et al. Evidence of mitochondrial dysfunction in fragile X-associated tremor/ataxia syndrome. *Biochem. J.* 2010; 429: 545-552.

[51] Cao Z, Hulsizer S, Tassone F, et al. Clustered burst firing in FMR1 premutation hippocampal neurons: amelioration with allopregnanolone. *Hum. Mol. Genet*. 2012; 21: 2923-2935.

[52] Garcia-Arocena D, Yang JE, Brouwer JR, et al. Fibroblast phenotype in male carriers of FMR1 premutation alleles. *Hum. Mol. Genet*. 2010; 19: 299-312.

[53] Jenkins EC, Tassone F, Ye L, et al. Reduced telomere length in older men with premutation alleles of the fragile X mental retardation 1 gene. *Am. J. Med. Genet. A*. 2008; 146A: 1543-1546.

[54] Hagerman R, Hagerman P. Advances in clinical and molecular understanding of the FMR1 premutation and fragile X-associated tremor/ataxia syndrome. *Lancet Neurol*. 2013; 12: 786-798.

[55] Chen Y, Tassone F, Berman RF, et al. Murine hippocampal neurons expressing Fmr1 gene premutations show early developmental deficits and late degeneration. *Hum. Mol. Genet*. 2010; 19: 196-208.

[56] Cunningham CL, Martínez Cerdeño V, Navarro Porras E, et al. Premutation CGG-repeat expansion of the Fmr1 gene impairs mouse neocortical development. *Hum. Mol. Genet*. 2011; 20: 64-79.

[57] Qin M, Entezam A, Usdin K, et al. A mouse model of the fragile X premutation: effects on behavior, dendrite morphology, and regional rates of cerebral protein synthesis. *Neurobiol Dis*. 2011; 42: 85-98.

[58] Sellier C, Usdin K, Pastori C, et al. The multiple molecular facets of fragile X-associated tremor/ataxia syndrome. *J. Neurodev. Disord.* 2014; 6: 23.

[59] Todd PK, Oh SY, Krans A, et al. CGG repeat-associated translation mediates neurodegeneration in fragile X tremor ataxia syndrome. *Neuron.* 2013; 78: 440-455.

[60] Ladd PD, Smith LE, Rabaia NA, et al. An antisense transcript spanning the CGG repeat region of FMR1 is upregulated in premutation carriers but silenced in full mutation individuals. *Hum. Mol. Genet.* 2007; 16: 3174-3187.

[61] Khalil AM, Faghihi MA, Modarresi F, et al. A novel RNA transcript with antiapoptotic function is silenced in fragile X syndrome. *PLoS One.* 2008; 3: e1486.

[62] Pastori C, Peschansky VJ, Barbouth D, et al. Comprehensive analysis of the transcriptional landscape of the human FMR1 gene reveals two new long noncoding RNAs differentially expressed in Fragile X syndrome and Fragile X-associated tremor/ataxia syndrome. *Hum. Genet.* 2014; 133: 59-67.

In: Allelic Forms of the FMR1 Gene
Editor: Montserrat Milà
ISBN: 978-1-63321-914-4

Chapter 4

Psychiatric Aspects in Fragile X Syndrome and Related Phenotypes

***E. Mur*[1] *and M. Milà*[*2-4]**
[1]CSM Martí i Julià. Instituto de Neuropsiquiatría y Adicciones (INAD). Parc de Salut Mar. Barcelona. Spain
[2]Servicio de Bioquímica i Genética Molecular. Hospital Clínic. Barcelona. Spain
[3]CIBER de EnfermedadesRaras (CIBERER) Barcelona, Spain
[4]IDIBAPS (Institut de Investigacions Biomèdiques August Pi I Sunyer) Barcelona, Spain

Abstract

This chapter describes the psychopathological alterations of the different phenotypes associated with the *FMR1* gene. Fragile X patients present enormous functional impairment with a behavioral phenotype of hyperactivity and impaired attention, marked anxiety, with poor eye contact, affective liability, aggression, self-injurious behavior and autistic features. Since its first description the basic formulation of the fragile X behavioral phenotype has remained intact to the present day, with substantial confirmation of these basic findings in subsequent studies. Autism is one of the most recognized and severe behavioral abnormalities observed in males with fragile X syndrome (FXS) and this syndrome is the leading known monogenic cause of autism, accounting for approximately 5% of autism cases. Approximately 30–50% of FXS individuals meet full Diagnostic and Statistical Manual of Mental Disorders DSM-IV-TR criteria for autism with 60–74% fulfilling criteria for an autism spectrum disorder (ASD). Attention deficit and hyperactivity disorder (ADHD) is the most common diagnosable condition in FXS patients, with most males meeting formal criteria at some point in their lives. In children with FXS, ADHD is reported to be characterized by more inattentiveness, restlessness, fidgetiness and impulsivity. However, affective symptoms can be severe and disruptive, and are a common target for psychopharmacologic intervention. Several clinical subgroups present a higher risk for presenting anxiety outcomes, including children with FXS. Males with FXS display a broad range of anxiety symptoms, but these symptoms often do not fit into the established categories of major anxiety disorders employed by the DSM. Regarding individuals carrying a premutation there is a growing body of literature suggesting that neuropsychiatric features of fragile X-associated tremor/ataxia syndrome (FXTAS) follow a fronto-subcortical pattern with primary impairments in executive function and increased vulnerability to mood and anxiety disorders. Increased rates of psychiatric symptoms may represent early markers of neurodegenerative diseases and have also been reported among *FMR1* premutation carriers without FXTAS. On the other hand, several studies have reported an excess of intermediate *FMR1* alleles in patients with cognitive and/or behavioral phenotypes. Numerous studies have investigated neuropsychological phenotypes among premutation allele carriers in women, but no definitive profile has been achieved. An increased risk of anxiety and mood disorders among premutation allele carriers has not been established, although they seem to be more common in these subjects than in controls.

Keywords: ASD, *FMR1*, hyperactive disorders, mood disorders, anxiety disorders

Introduction

The fragile X Syndrome (FXS) is associated with a complex but relatively consistent psychiatric phenotype and provides the enticing possibility of understanding a complex neuropsychiatric disorder at the synaptic, molecular, and genetic levels. This chapter describes the psychopathological alterations of the different phenotypes associated with the different status of the *FMR1* gene: FXS (>200 CGG), premutated individuals (55-200 CGG) and intermediate alleles (45-54 CGG). In 1984, Fryns [1] studied a large population of fragile X males, describing a behavioral phenotype of hyperactivity and impaired attention, marked anxiety with poor eye contact, affective liability, aggression, self-injurious behavior (especially the characteristic hand biting), and autistic features reported as repetitive, perseverative and stereotypic behaviors. This basic formulation of the fragile X behavioral phenotype has remained intact to the present day, with substantial confirmation of these basic findings in subsequent studies.

Fragile X Syndrome

Fragile X patients suffer maladaptive behaviors and emotional disturbance with an enormous functional impairment; symptoms are a frequent reason for families to seek treatment and can lead to institutionalization in more severe cases.

Autism Disorders

Autistic disorder is the most debilitating subgroup of a larger category known as pervasive developmental disorders (American Psychiatric Association) characterized by impairment in social interaction and verbal and non-verbal communication, and restricted, repetitive and stereotypic patterns of behavior, interests and activities. Although there is considerable variability in individual symptoms, core deficits in social communication and restricted and repetitive behaviors are hallmarks of the disorder [2]. FXS is the leading known monogenic cause of autism, accounting for approximately 5% of autism cases [reviewed in 3] and autism is one of the most recognized and severe behavioral abnormalities observed in males with FXS [4-8]. In individuals with the full mutation, approximately 30–50% meet full DSM-IV-TR criteria for autism with 60–74% fulfilling criteria for an autism spectrum disorder (ASD) [9-13]. Over 90% of individuals with FXS display some form of atypical behavior characteristic of autism, including social interaction (e.g. avoidance of eye contact, social withdrawal, and social anxiety) and repetitive and stereotyped behaviors [14]. Co-morbid FXS and autism are indicative of worse developmental outcomes [15-17], and greater impairment in cognition and adaptive behavior skills are more severe aberrant behavior than FXS without autism [18]. It has been suggested that autistic behaviors increase slowly but significantly over time, as do associated social avoidance behaviors [19].

The main dilemma arises when considering the pathogenesis of clinical autism spectrum in patients diagnosed with the FXS. Some authors consider that the clinical autism spectrum in these patients is a consequence of pathophysiological alterations resulting from the mutation of the *FMR1* gene, while other authors consider that this clinical autism spectrum in these patients has an etiopathological mechanism similar to idiopathic autism. These different points of view are generated due to the fact that diagnostic criteria for autism spectrum disorders are purely clinical.

Likewise, Harris (2011) [20] has proposed a focus on brain-behavior relationships targeting advancement of behavioral phenotyping in neurogenetic disorders precluding the application of DSM-IV diagnostic behavioral criteria to identify disorders such as FXS. He proposed that FXS is a neural model and phenocopy of autism and should not be considered a genetic model for autism. On the other hand, a number of investigators have reported findings indicating highly similar profiles between individuals with FXS and autism versus idiopathic autism and apply categorical or dimensional ratings of autism in FXS [9,16,21-23]. These authors agree with the concept that the autism phenomenon represents a range of behaviors, and perhaps also of other neurologic features [24]. While there is clear consensus regarding the shared phenomenology between FXS and idiopathic autism, there is great debate regarding diagnostic issues. Two of the primary debates on FXS center around questions of whether autism in FXS represents a continuum, with only those most severely affected meeting criteria for autism, and whether autism in FXS is the same as or different from idiopathic autism [9]. Although the literature on the co-morbidity of FXS and autism is extensive, few published studies have longitudinally examined early indicators of autism in infants. Study of the emerging characteristics in FXS is critical for understanding if early features in infants with FXS are associated with later autistic behaviors as reported in idiopathic autism. To date, the studies that have been conducted lend evidence to the fact that indicators of autism are present at as early as 12 months of age in males with FXS, and replicate findings in idiopathic autism that implicate difficulties, with disengagement of attention and shifts in behavior at 6 to 12 months of age in the later development of autism [25-27]. Clearly, brain–behavior relationships in idiopathic autism and FXS are complex, and there is currently insufficient evidence to resolve these debates. In summary, however, despite all the controversies generated in the diagnosis of clinical ASD, at present the *FMR1* gene is considered the most common monogenic cause of ASD, and therefore, the molecular diagnosis of FXS should be ruled out in all ASD [28].

Hyperactivity Disorders

Attention deficit and hyperactivity (ADHD) is the most common diagnosable condition in FXS patients, with most males meeting formal criteria at some point in their lives. This condition is typically not stable over time in any given individual [1]. Using DSM-V, several of the ADHD symptoms must be present in individuals prior to the age of 12 years, compared to 7 years as the age of onset in DSM-IV. This change is supported by substantial research published since 1994 that found no clinical differences between children identified at 7 years versus later in terms of the course, severity, outcome, or treatment response. DSM-5 includes no exclusion criteria for people with ASD, since symptoms of both disorders co-occur. The higher inattentiveness, restlessness, fidgetiness and impulsivity in ADHD in children with FXS is suggestive of the ADHD inattentive sub-type compared to ADHD in the general population; these features do not necessarily improve with age [29]. The signature of the FXS is strong unsustained attention, but poor capacity to switch between tasks and weaker inhibitory control [reviewed in 30].

Very young children with FXS are often noted to be physically hypoactive, with somewhat impaired attention. Preschool children can display dramatic increases in activity levels, leading to markedly disruptive behavior. As children grow, hyperactivity declines with increasing body mass, while problems with attention continue throughout life. This can be seen as similar to the course of ADHD in the normal population, though the degree of hyperactivity in FXS is impressive. There is also evidence that the attention deficit seen in males with FXS has a specific profile [31], which is distinct from other causes of developmental disorders, suggesting that the attention problems seen in the course of FXS may represent more than nonspecific immaturity. There is evidence that ADHD symptoms in FXS respond to stimulants [32,33].

Mood Disorders

There has been considerable controversy regarding the behavior phenotype of FXS and the nature of behavioral and emotional problems associated with it.

As described by Backes and colleagues (2000) [34], males with FXS rarely meet formal criteria for a diagnosis of a major mood disorder as defined in DSM-V. Diagnoses such as major depression or bipolar disorder require periods of abnormal mood that are sustained, whereas individuals with FXS typically exhibit labile mood, irritability, self-injurious behavior, and aggressive outbursts of a more fleeting and episodic nature, not meeting the conventional duration criteria. These episodes are typically adaptive, related to environmental stressors and are less frequent in familial or more structured settings. However, affective symptoms can be severe and disruptive, and psychopharmacologic intervention is needed.

Selective serotonin reuptake inhibitors are a commonly employed treatment strategy for affective symptoms, along with other antidepressants, anticonvulsants, and atypical antipsychotics in more severe cases [33]. There have been no clinical trials of any size, open or controlled, of antidepressants or anticonvulsants for the treatment of affective symptoms of FXS. Curiously, the study of valproic acid by Torrioli and collaborators (2010) [35] focused exclusively on ADHD symptoms in boys and they considered that this treatment could be considered as an alternative to treating symptoms with stimulants the efficacy of which needs to be confirmed by further studies in patients with FXS. Behavioral problems in FXS improve from childhood to young adulthood [36].

Anxiety Disorders

Children with neurodevelopmental disorders, such as FXS, have a higher risk of presenting anxiety. Patients with FXS display a broad range of anxiety symptoms, but these symptoms often do not fit into the established categories of major anxiety disorders employed by the DSM. Cordeiro and coworkers (2011) [37] examined the prevalence of anxiety disorders in a FXS population using DSM-IV-TR criteria. They found that 83 % of participants (n=97, ages 5.5–33.3 years) met criteria for any anxiety disorder, and 58 % exhibited multiple anxiety disorders. In addition to the high proportion of anxiety disorders in FXS, an important part of individuals with FXS display autistic symptoms as mentioned previously.

Symptoms of anxiety and autism overlap and interrelate within FXS, although these disorders have been distinguished through behavioral and physiological profiles. Kaufmann, Budimirovic and colleagues have published a series of papers characterizing anxiety and autism as distinct social interaction disorders in FXS [38-39]. These authors propose that social anxiety emerges from a combination of lower non-verbal abilities and moderate social withdrawal, whereas autism (either alone or in conjunction with social anxiety) is characterized by a more complex constellation of severe social withdrawal and lower adaptive socialization or verbal skills.

Negative affect is one of the most commonly studied predictors of problem behaviors in non-clinical [40-41] and clinical [42] populations. Negative affect is composed of several specific dimensions of temperament; including fear, approach, soothability, sadness, anger, discomfort, and motor activity [43] and predicts anxiety in preschool boys with FXS [44]. Multilevel models indicate associations between elevated anxiety and higher fear and sadness, lower soothability, and steeper longitudinal increases in approaches in the FXS population.

A minority of males with FXS fulfill formal criteria for the diagnosis of obsessive-compulsive disorder (OCD), while "compulsive symptoms" have been noted in several studies in a large majority of subjects with FXS. In most cases of FXS, individuals exhibit symptoms strongly reminiscent of obsessions and compulsions, but which do not meet the precise psychiatric definitions for these symptoms. Often, pleasure is derived from repetitive and "compulsive" behaviors, in contrast to the ego-dystonic nature of true obsessions and compulsions. Hoarding, counting, and the need for symmetry are all typical symptoms of OCD frequently seen in FXS. Similarly, younger children with FXS meet the criteria for separation anxiety disorder in a small minority of cases [34], while symptoms of separation anxiety, social phobia, panic, and agoraphobia are seen clinically at a much higher rate.

The psychiatric drugs currently available can provide significant symptomatic relief of the hyperactivity, anxiety disorders, and affective disturbances often seen in the course of FXS. However, patients with this syndrome may be especially susceptible to the psychiatric side effects of these medications, requiring particular care in their prescription.

Premutation Carriers

For many years, premutation carriers of the *FMR1* gene were considered asymptomatic, but various clinical disorders have been associated in both men and women, being Fragile X tremor –ataxia syndrome (FXTAS) one of the most severe. Individuals with carrier premutation are those that have alleles between 55-200 CGG. Some of the clinical features have been related to the CGG number; in general, more length in the CGG track implies more severe clinical manifestations.

There is a growing body of evidence suggesting that *FMR1* premutation carriers may have increased vulnerability of presenting psychiatric disorders [45]. However, the presence of neuropsychological and behavioral impairment among *FMR1* premutation carriers remains controversial, as these features were initially thought to be associated with the stress of raising children with FXS [reviewed in 46].

Altered neurobehavioral profiles including variation of phenotypes associated with mood and anxiety may be expected among younger premutation carriers.

An increased risk of anxiety and mood disorders among premutation carriers has not been established. Some studies have reported a lack of phenotype [47-48], while others have described repeat length associations with psychiatric symptoms [49-50]. Hessl and coworkers (2005)[51] found that the *FMR1* transcript level, but not repeat length or Fragile X Mental retardation Protein (FMRP) levels, was significantly associated with increased severity of psychiatric symptoms in males, independently of FXTAS status. These results suggest that premutation carriers may be at risk of presenting emotional morbidity; however, phenotypic differences were subtle and of a small CGG effect size.

The largest and most recent study of life-time mood and anxiety in the premutation population was completed by Bourgeois and colleagues (2009) [52]. In this study, the prevalence of anxiety disorders in carriers with and without FXTAS was compared with a very large age-matched national dataset. In terms of all anxiety disorders, only those with FXTAS demonstrated a higher prevalence. Upon separation, this was similarly true for panic disorder, post-traumatic stress disorder and specific phobia. Generalized anxiety disorder and OCD failed to demonstrate any difference between carriers and controls. Only social phobia was found to have higher levels in premutation carriers without FXTAS compared to controls. Chronic anxiety has also been associated with radiological signs on MRI; specifically, the higher the anxiety score the smaller the size of the hippocampus in women with the premutation [53].

Rodriguez-Revenga and collaborators (2008) [54] examined psychiatric and depressive symptoms in 34 *FMR1* premutation carrier mothers of children with FXS in comparison with two control groups (39 mothers with a non-FXS intellectual disability child and 39 mothers from the general population). Both groups of mothers with a child with intellectual disability showed greater susceptibility to psychological problems than the control group without a mentally retarded child, but *FMR1* premutated mothers evidenced a higher tendency to depression. These results suggest that, despite the stress of caring for a child with mental retardation, the premutation by itself could be responsible for some psychiatric traits.

In a screening study of individuals from families with FXS, roughly 14% of boys and 5% of girls with the premutation were found to also have an ASD [10]. Even among those carriers not diagnosed with ASD, related psychological traits are more common among carriers compared to controls without the premutation. A recent study examined a broad range of pragmatic language skills as well as related behavioral features of the broad autism phenotype among women with the premutation compared with mothers of children with autism, and mothers of typically developing children with no family history of FXS, autism, or language impairment [55]. In this study, conversational samples from a semistructured videotaped interview were used to assess pragmatic language using the Pragmatic Rating Scale (PRS) [56]. This study replicated previous findings in the autism parent group and also showed that women with the premutation exhibited similarly elevated rates of pragmatic language problems relative to the controls. The presence of broad autism phenotype traits was associated with greater expression of autism symptoms in their children with FXS. Other studies have also found increased rates of both social aloofness [49] and a rigid perfectionism [57] among carrier women.

Given its relative rarity in the general population, psychosis has been challenging to study in premutation carriers. Initial linkage analysis failed to show a clear relationship of schizophrenia to the *FMR1* gene [58]. Prevalence studies have found the overall rate of psychotic disorders to be low [49]. There have, however, been several case reports of combined psychotic illnesses and the premutation, including schizoaffective disorder [59] and with combined schizophrenia and schizoid personality disorder [60]. Interestingly, as opposed to frank psychotic disorders, multiple studies have found an increased prevalence of schizotypal personality traits in the carrier population [49,61].

Attention regulation difficulties have been proposed to be a problem in people with the premutation. Notably, when compared with their control siblings, premutation carriers had significantly more issues related to attention than their noncarrier siblings [62]. Inattention and impulsivity amongst *FMR1* carriers can be problematic through adulthood [63], although hyperactivity was not noted to be increased in prevalence.

Dysthymia and bipolar disorder have generally failed to demonstrate significant levels in carriers compared to controls [52].

Fragile X Associated Tremor Ataxia Syndrome (FXTAS)

FXTAS is an X-linked neurodegenerative disorder affecting up to 45.5% of males and 16.5 % of females carrying a premutation in the *FMR1* gene [64-65].

There is a growing body of literature suggesting that the neuropsychiatric features of FXTAS follow a fronto-subcortical pattern with primary impairments in executive function and increased vulnerability to mood and anxiety disorders [51,53, 66-69]

Increased rates of psychiatric symptoms may represent early markers of neurodegenerative diseases such as Parkinsons Disease (PD) [70] and Huntington's Disease [71], which have also been reported among *FMR1* premutation carriers without FXTAS, including obsessive-compulsive symptoms [51], social phobia [72], depression [50], schizotypal features [73] and abnormalities in social cognition [74].

Three studies have examined the psychiatric features of FXTAS: (1) Bacalman et al. (2006) [67]; (2) Adams et al. (2010) [53]; and (3) Hashimoto et al. (2011) [69]. They showed no significant association between FXTAS and different psychiatric diagnoses. The studies compared the prevalence of neuropsychiatric symptoms between individuals with FXTAS and matched controls with normal *FMR1* alleles (cohort 1 and 3), and between asymptomatic *FMR1* premutation carriers in cohort 2.

Symptoms of depression were significantly elevated among males with FXTAS when measured using the informant-rated NPI in cohort 1; however, self-reported symptoms measured using the SCL-90-R in cohort 3 were comparable to controls. Women with FXTAS (cohort 2) tended to report higher rates of depression than controls with normal *FMR1* alleles; however, this difference did not withstand correction for multiple comparisons.

Anxiety was significantly elevated among older males with FXTAS compared to controls in cohort 2; however, in cohort 1 the group difference (50% FXTAS group vs. 0% control group) was not significant, possibly due to the small sample size (n=14). Male carriers with FXTAS exhibited elevated rates of anxiety in cohort 3 (Cohen's d=0.69) and comparable scores in cohort 2. Females with FXTAS (cohort 2) exhibited higher rates of anxiety and obsessive-compulsive symptoms compared to controls with normal *FMR1* alleles and asymptomatic *FMR1*premutation carriers.

Males with FXTAS (cohorts 2 and 3) reported significantly higher rates of obsessive compulsive symptoms compared to asymptomatic carriers (Cohen's d=0.92, cohort 7), but not controls (cohort 2). Symptoms of anxiety among asymptomatic male and female *FMR1* premutation carriers were compared to controls with normal alleles in cohort 2 and no group differences were detected.

Additional informant-rated behavioral disturbances found to be significantly elevated among males with FXTAS (cohort 1) compared to controls included apathy, irritability, disinhibition, and agitation/aggression (Cohen's d=3.53) [67].

Increased obsessive-compulsive and depressive symptoms (but not anxiety) were also associated with decreased left amygdala volume among males with FXTAS in cohort 3 (all ps<0.02) [69].

It is difficult to determine the psychiatric features most commonly associated with FXTAS based on this limited number of studies; however, the evidence available suggests that FXTAS may be associated with increased rates of depression, anxiety and obsessive-compulsive symptoms. This is supported by a previous systematic review that included case reports and case series, suggesting elevated rates of mood and anxiety disorders among those with FXTAS [52].

Emerging evidence suggests that a subset of carriers without FXTAS may exhibit cognitive deficits, psychiatric symptoms, and changes in brain structure and function which may be modulated by *FMR1* gene expansion [51,53,69,74-75]. Specifically, abnormalities in structural and functional connectivity have been described in the middle cerebellar peduncles, hippocampus, amygdala and prefrontal cortex, raising the possibility that these early indicators of fronto-subcortical and cortico-cerebellar dysfunction may be detectable prior to the onset of clinical symptoms. The capacity to identify the subjects most at-risk or closest to onset of illness would be invaluable for recruitment into clinical trials as treatments for FXTAS become available. In the absence of longitudinal follow-up of these cohorts, however, it is unclear whether cognitive, psychiatric and radiological changes observed in some asymptomatic carriers represent an independent developmental phenotype associated with the *FMR1* premutation, or serve as early indicators of later progression to FXTAS.

Intermediate *FMR1* alleles (alleles within the 45–54 CGG repeat range are described as an 'intermediate or gray zone").During the last few years, several studies have reported an excess of intermediate *FMR1* alleles in patients with cognitive and/or behavioral phenotypes [76-80] while other studies do not support this relationship. Madrigal and colleagues (2011) [81] did not found evidence of an association between intermediate alleles and behavioral or cognitive phenotypes, suggesting that intermediate alleles are not clearly implicated in these pathologies.

A frequent concern in FXS screening is the genetic counseling to intermediate allele carriers. Currently, the follow-up of individuals with no AGG interruptions is indicated, despite the low risk of expansion in the next generation. Intermediate alleles should not be considered a risk factor for ID or behavioral phenotypes. Nonetheless, these alleles should be characterized to provide accurate genetic counseling [81].

Table 1. Percentage of affected individuals for the psychiatric aspects

Psychiatric features	FXS	PM with FXTAS	PM without FXTAS	References
Autism traits	>90%			[14]
Autism criteria DSM-IV	30-50%			[11-13]
Autism spectrum disorders	60-74%		14% Males 5% females	[11-12,57]
ADHD	30% Females 60% Males		30% Females 14% Males	[62]
Affective disorders	22% Females 12% Males	65%	55.7% 34.2%	[49,62,72]
Anxiety	83%	52%	35-40%	[37,49,72,]

Conclusion

FMR1 is the first monogenic cause of autism disorder. Approximately 30–50% of FXS meet full DSM-IV-TR criteria for autism, 60–74% fulfilling criteria for an ASD and over 90% of individuals display some form of atypical behavior characteristic of autism. Molecular diagnosis of FXS should be ruled out in all ASD. Increased rates of psychiatric symptoms may represent early markers of neurodegenerative diseases which have been reported among *FMR1* premutation carriers with and without FXTAS; including obsessive-compulsive symptoms, social phobia, depression, schizotypal features, and abnormalities in social cognition. Altered neurobehavioral profiles including a variation of phenotypes associated with mood and anxiety may be expected among younger premutation carriers. The trend is to present higher rates of anxiety and mood disorders in premutation carriers, but no study has found significant differences between carriers and the control population.

References

[1] Fryns JP. The fragile X syndrome. A study of 83 families. *Clin. Genet.* 1984;26(6):497–528.

[2] Tager-Flusberg H, Rogers S, Cooper J, Landa R, Lord C, Paul R, Rice M, Stoel-Gammon C, Wetherby A, Yoder P. Defining spoken language benchmarks and selecting measures of expressive language development for young children with autism spectrum disorders. J Speech Lang Hear Res. 2009 Jun;52(3):643-52. Hagerman RJ, Hall DA, Coffey S, Leehey M, Bourgeois J, Gould J, Zhang L, Seritan A, Berry-Kravis E, Olichney J, Miller JW, Fong AL, Carpenter R, Bodine C, Gane LW, Rainin E, Hagerman H, Hagerman PJ. Treatment of fragile X-associated tremor ataxia syndrome (FXTAS) and related neurological problems. *Clin. Interv Aging.* 2008;3(2):251-62. Review.

[3] Hagerman RJ, JacksonAW III, Levitas A, Rimland B, Barden M, 1986. An analysis of autism in fifty males with the fragile X syndrome. *Am. J. Med. Genet* 23:359-374.

[4] Baumgardner TL, Reiss AL, Freund LS, Abrams MT. Specification of the neurobehavioral phenotype in males with fragile X syndrome. *Pediatrics* 1995;95(5):744–752.

[5] Bailey DB Jr, Mesibov GB, Hatton DD, Clark RD, Roberts JE, Mayhew L. Autistic behavior in young boys with fragile X syndrome. *J. Autism Dev. Disord*. 1998 Dec;28(6):499-508.

[6] Kaufman WE, Reiss AL. Molecular and cellular genetics of fragile X syndrome. *Am. J. Med. Genet*.1999;88:11-24.

[7] Hagerman RJ. The physical and behavioral phenotype. In Hagerman RJ, Hagerman PJ, editors. Fragile X syndrome: Diagnosis, treatment, and research. Baltimore, MD: The Johns Hopkins Univerity Pres. 2002;pp 3-109

[8] Kaufmann WE, Cortell R, Kau AS, Bukelis I, Tierney E, Gray RM, Cox C, Capone GT, Stanard P. Autism spectrum disorder in fragile X syndrome: communication, social interaction, and specific behaviors. *Am. J. Med. Genet A*. 2004 Sep 1;129A(3):225-34.

[9] Clifford S, Dissanayake C, Bui QM, Huggins R, Taylor AK, Loesch DZ. Autism spectrum phenotype in males and females with fragile X full mutation and premutation. *J. Autism Dev. Disord* 2007, 37:738–747.

[10] Hall SS, Lightbody AA, Reiss AL. Compulsive, self-injurious, and autistic behavior in children and adolescents with fragile X syndrome. *Am. J. Ment. Retard*. 2008 Jan;113(1):44-53.

[11] Harris SW, Hessl D, Goodlin-Jones B, Ferranti J, Bacalman S, Barbato I, Tassone F, Hagerman PJ, Herman H, Hagerman RJ. Autism profiles of males with fragile X syndrome. *Am. J. Ment. Retard*. 2008 Nov;113(6):427-38.

[12] Bailey DB, Raspa M, Holiday D, Bishop E, Olmsted M. Functional skills of individuals with fragile x syndrome: a lifespan cross-sectional analysis. *Am. J. Intellect Dev. Disabil*. 2009 Jul;114(4):289-303.

[13] Hernandez RN, Feinberg RL, Vaurio R, Passanante NM, Thompson RE, Kaufmann WE. Autism spectrum disorder in fragile X syndrome: a longitudinal evaluation. *Am. J. Med. Genet A*. 2009 Jun;149A(6):1125-37.

[14] Bailey DB Jr, Hatton DD, Mesibov G, Ament N, Skinner M. Early development, temperament, and functional impairment in autism and fragile X syndrome. *J. Autism Dev. Disord*. 2000 Feb;30(1):49-59.

[15] Rogers SJ, Wehner DE, Hagerman R. The behavioral phenotype in fragile X: symptoms of autism in very young children with fragile X syndrome, idiopathic autism, and other developmental disorders. *J. Dev. Behav. Pediatr.* 2001 Dec;22(6):409-17.

[16] Kau AS, Tierney E, Bukelis I, Stump MH, Kates WR, Trescher WH, Kaufmann WE. Social behavior profile in young males with fragile X syndrome: characteristics and specificity. *Am. J. Med. Genet A*. 2004 Apr 1;126A(1):9-17.

[17] Hatton DD, Sideris J, Skinner M, Mankowski J, Bailey DB Jr, Roberts J, Mirrett P. Autistic behavior in children with fragile X syndrome: prevalence, stability, and the impact of FMRP. *Am. J. Med. Genet A*. 2006 Sep 1;140A(17):1804-13.

[18] Roberts JE, Weisenfeld LA, Hatton DD, Heath M, Kaufmann WE. Social approach and autistic behavior in children with fragile X syndrome. *J. Autism Dev. Disord*. 2007 Oct;37(9):1748-60. Epub 2006 Dec 19.

[19] Harris JC. Brain and behavior in fragile x syndrome and idiopathic autism. *Arch. Gen. Psychiatry*. 2011 Mar;68(3):230-1.

[20] Bailey DB Jr. Newborn screening for fragile X syndrome. *Ment. Retard Dev. Disabil Res. Rev*. 2004;10(1):3-10. Review.

[21] Lewis P, Abbeduto L, Murphy M, Richmond E, Giles N, Bruno L, Schroeder S. Cognitive, language and social-cognitive skills of individuals with fragile X syndrome with and without autism. *J. Intellect Disabil Res.* 2006 Jul;50(Pt 7):532-45.

[22] Hoeft F, Walter E, Lightbody AA, Hazlett HC, Chang C, Piven J, Reiss AL. Neuroanatomical differences in toddler boys with fragile x syndrome and idiopathic autism. *Arch. Gen. Psychiatry*. 2011 Mar;68(3):295-305. doi: 10.1001/archgenpsychiatry.2010.153. Epub 2010 Nov 1.

[23] Lord C, Petkova E, Hus V, Gan W, Lu F, Martin DM, Ousley O, Guy L, Bernier R, Gerdts J, Algermissen M, Whitaker A, Sutcliffe JS, Warren Z, Klin A, Saulnier C, Hanson E, Hundley R, Piggot J, Fombonne E, Steiman M, Miles J, Kanne SM, Goin-Kochel RP, Peters SU, Cook EH, Guter S, Tjernagel J, Green-Snyder LA, Bishop

S, Esler A, Gotham K, Luyster R, Miller F, Olson J, Richler J, Risi S. A multisite study of the clinical diagnosis of different autism spectrum disorders. *Arch. Gen. Psychiatry*. 2012 Mar;69(3):306-13. Epub 2011 Nov 7.

[24] Zwaigenbaum L, Bryson S, Rogers T, Roberts W, Brian J, Szatmari P. Behavioral manifestations of autism in the first year of life. *Int. J. Dev. Neurosci.* 2005 Apr-May;23(2-3):143-52.

[25] Baranek GT, Roberts JE, David FJ, Sideris J, Mirrett PL, Hatton DD, Bailey DB Jr. Developmental trajectories and correlates of sensory processing in young boys with fragile X syndrome. *Phys. Occup. Ther Pediatr.* 2008;28(1):79-98.

[26] Roberts TP, Cannon KM, Tavabi K, Blaskey L, Khan SY, Monroe JF, Qasmieh S, Levy SE, Edgar JC. Auditory magnetic mismatch field latency: a biomarker for language impairment in autism. *Biol. Psychiatry.* 2011 Aug 1;70(3):263-9. Epub 2011 Mar 9.

[27] Shen Y, Dies KA, Holm IA, Bridgemohan C, Sobeih MM, Caronna EB, Miller KJ, Frazier JA, Silverstein I, Picker J, Weissman L, Raffalli P, Jeste S, Demmer LA, Peters HK, Brewster SJ, Kowalczyk SJ, Rosen-Sheidley B, McGowan C, Duda AW 3rd, Lincoln SA, Lowe KR, Schonwald A, Robbins M, Hisama F, Wolff R, Becker R, Nasir R, Urion DK, Milunsky JM, Rappaport L, Gusella JF, Walsh CA, Wu BL, Miller DT; Autism Consortium Clinical Genetics/DNA Diagnostics Collaboration. Clinical genetic testing for patients with autism spectrum disorders. *Pediatrics.* 2010 Apr;125(4):e727-35.

[28] Turk J, Cornish K. Face recognition and emotion perception in boys with fragile-X syndrome. *J. Intellect Disabil. Res.* 1998 Dec;42 (Pt 6):490-9.

[29] Cornish KM, Turk J, Wilding J, Sudhalter V, Munir F, Kooy F, Hagerman R. Annotation: Deconstructing the attention deficit in fragile X syndrome: a developmental neuropsychological approach. *J. Child Psychol. Psychiatry.* 2004 Sep;45(6):1042-53. Review.

[30] Munir F, Cornish KM, Wilding J. A neuropsychological profile of attention deficits in young males with fragile X syndrome. *Neuropsychologia.* 2000;38(9):1261-70.

[31] Berry-Kravis E, Potanos K. Psychopharmacology in fragile X syndrome-present and future. *Ment. Retard Dev. Disabil Res. Rev.* 2004;10(1):42-8.

[32] Tsiouris JA, Brown WT. Neuropsychiatric symptoms of fragile X syndrome: pathophysiology and pharmacotherapy. *CNS Drugs.* 2004;18(11):687-703.
[33] Backes M, Genç B, Schreck J, Doerfler W, Lehmkuhl G, von Gontard A. Cognitive and behavioral profile of fragile X boys: correlations to molecular data. *Am. J. Med. Genet.* 2000 Nov 13;95(2):150-6.
[34] Torrioli M, Vernacotola S, Setini C, Bevilacqua F, Martinelli D, Snape M, Hutchison JA, Di Raimo FR, Tabolacci E, Neri G. Treatment with valproic acid ameliorates ADHD symptoms in fragile X syndrome boys. *Am. J. Med. Genet A.* 2010 Jun;152A(6):1420-7.
[35] Einfeld S, Tonge B, Turner G. Longitudinal course of behavioral and emotional problems in fragile X syndrome. *Am. J. Med. Genet.* 1999 Dec 22;87(5):436-9.
[36] Cordeiro L, Ballinger E, Hagerman R, Hessl D. Clinical assessment of DSM-IV anxiety disorders in fragile X syndrome: prevalence and characterization. *J. Neurodev Disord.* 2011 Mar;3(1):57-67.
[37] Budimirovic DB, Bukelis I, Cox C, Gray RM, Tierney E, Kaufmann WE. Autism spectrum disorder in Fragile X syndrome: differential contribution of adaptive socialization and social withdrawal. *Am. J. Med. Genet A.* 2006 Sep 1;140A(17):1814-26.
[38] Budimirovic DB, Kaufmann WE. What can we learn about autism from studying fragile X syndrome? *Dev. Neurosci.* 2011;33(5):379-94.. Review.
[39] Calkins SD, Blandon AY, Williford AP, Keane SP. Biological, behavioral, and relational levels of resilience in the context of risk for early childhood behavior problems. *Dev. Psychopathol.* 2007 Summer;19(3):675-700.
[40] Lemery KS, Essex MJ, Smider NA. Revealing the relation between temperament and behavior problem symptoms by eliminating measurement confounding: expert ratings and factor analyses. *Child Dev.* 2002 May-Jun;73(3):867-82.
[41] Hutman T, Rozga A, DeLaurentis AD, Barnwell JM, Sugar CA, Sigman M. Response to distress in infants at risk for autism: a prospective longitudinal study. *J. Child Psychol. Psychiatry.* 2010 Sep;51(9):1010-20.

[42] Putnam SP, Stifter CA. Behavioral approach-inhibition in toddlers: prediction from infancy, positive and negative affective components, and relations with behavior problems. *Child Dev.* 2005 Jan-Feb;76(1):212-26.

[43] Tonnsen BL, Malone PS, Hatton DD, Roberts JE. Early negative affect predicts anxiety, not autism, in preschool boys with fragile X syndrome. *J. Abnorm Child Psychol*. 2013 Feb;41(2):267-80.

[44] Hunter JE, Leslie M, Novak G, Hamilton D, Shubeck L, Charen K, Abramowitz A, Epstein MP, Lori A, Binder E, Cubells JF, Sherman SL. Depression and anxiety symptoms among women who carry the FMR1 premutation: impact of raising a child with fragile X syndrome is moderated by CRHR1 polymorphisms. *Am. J. Med. Genet B Neuropsychiatr Genet*. 2012. Epub 2012 May 9.

[45] Hunter JE, Abramowitz A, Rusin M, Sherman SL. Is there evidence for neuropsychological and neurobehavioral phenotypes among adults without FXTAS who carry the FMR1 premutation? A review of current literature. *Genet Med*. 2009 Feb;11(2):79-89. Review.

[46] Reiss AL, Freund L, Abrams MT, Boehm C, Kazazian H. Neurobehavioral effects of the fragile X premutation in adult women: a controlled study. *Am. J. Hum. Genet* 1993, 52:884–894.

[47] Sobesky WE, Taylor AK, Pennington BF, Bennetto L, Porter D, Riddle J, Hagerman RJ. Molecular/clinical correlations in females with fragile X. *Am. J. Med. Genet*. 1996 Aug 9;64(2):340-5.

[48] Franke P, Leboyer M, Gansicke M, Weiffenbach O, Biancalana V, Cornillet-Lefebre P, Maier W. Genotype-phenotype relationship in female carriers of the premutation and full mutation of FMR-1. *Psychiatry Res* 1998, 80:113–127.

[49] Johnston C, Eliez S, Dyer-Friedman J, Hessl D, Glaser B, Blasey C, Taylor A, Reiss A. Neurobehavioral phenotype in carriers of the fragile X premutation. *Am. J. Med. Genet*. 2001 Nov 1;103(4):314-9.

[50] Hessl D, Tassone F, Loesch DZ, Berry-Kravis E, Leehey MA, Gane LW, Barbato I, Rice C, Gould E, Hall DA, Grigsby J, Wegelin JA, Harris S, Lewin F, Weinberg D, Hagerman PJ, Hagerman RJ. Abnormal elevation of FMR1 mRNA is associated with psychological symptoms in individuals with the fragile X premutation. *Am. J. Med. Genet B Neuropsychiatr Genet*. 2005 Nov 5;139B(1):115-21.

[51] Bourgeois JA, Coffey SM, Rivera SM, Hessl D, Gane LW, Tassone F, Greco C, Finucane B, Nelson L, Berry-Kravis E, Grigsby J, Hagerman PJ, Hagerman RJ. A review of fragile X premutation disorders: Expanding the psychiatric perspective. *J. Clin. Psychiatry* 2009, 70:852–862.

[52] Adams PE, Adams JS, Nguyen DV, Hessl D, Brunberg JA, Tassone F, Zhang W, Koldewyn K, Rivera SM, Grigsby J, Zhang L, Decarli C, Hagerman PJ, Hagerman RJ. Psychological symptoms correlate with reduced hippocampal volume in fragile X premutation carriers. *Am. J. Med. Genet B Neuropsychiatr Genet.* 2010 Apr 5;153B(3):775-85.

[53] Rodriguez-Revenga L, Madrigal I, Alegret M, Santos M, Milà M. Evidence of depressive symptoms in fragile-X syndrome premutated females. *Psychiatr Genet.* 2008 Aug;18(4):153-5.

[54] Losh M, Klusek J, Martin GE, Sideris J, Parlier M, Piven J. Defining genetically meaningful language and personality traits in relatives of individuals with fragile X syndrome and relatives of individuals with autism. *Am. J. Med. Genet B* 2012, 159B:660–668.

[55] Landa R, Piven J, Wzorek MM, Gayle JO, Chase GA, Folstein SE: Social language use in parents of autistic individuals. *Psychol. Med.* 1992, 22:245–254

[56] Hessl D, Rivera SM, Reiss AL: The neuroanatomy and neuroendocrinology of fragile X syndrome. *Ment. Retard Dev.* D R 2004, 10:17–24

[57] Ashworth A, Abusaad I, Walsh C, Nanko S, Murray RM, Asherson P, Collier DA. Linkage analysis of the fragile X gene FMR-1 and schizophrenia: no evidence for linkage but report of a family with schizophrenia and an unstable triplet repeat. *Psychiatr Genet* 1996, 6:81–86.

[58] Al-Semaan Y, Malla AK, Lazosky A: Schizoaffective disorder in a fragile-X carrier. *Aust. N Z J. Psychiatry* 1999, 33:436–440.

[59] Khin NA, Tarleton J, Raghu B, Park SK: Clinical description of an adult male with psychosis who showed FMR1 gene methylation mosaicism. *Am. J. Med. Genet* 1998, 81:222–224.

[60] Thompson NM, Rogeness GA, McClure E, Clayton R, Johnson C: Influence of depression on cognitive functioning in Fragile X females. *Psychiatry Res* 1996, 64:97–104.

[61] Bailey DB Jr, Raspa M, Olmsted M, Holiday DB. Co-occurring conditions associated with FMR1 gene variations: findings from a national parent survey. *Am. J. Med. Genet A*. 2008 Aug 15;146A(16):2060-9.
[62] Hunter JE, Allen EG, Abramowitz A, Rusin M, Leslie M, Novak G, Sherman SL. No evidence for a difference in neuropsychological profile among carriers and noncarriers of the FMR1 premutation in adults under the age of 50. *Am. J. Hum. Genet.* 2008, 83:692–702.
[63] Hagerman RJ, Hagerman PJ.Fragile X syndrome: a model of gene-brain-behaviour relationships. *Rev. Neurol.* 2001 Oct;33 Suppl 1:S51-7. Review. Spanish.
[64] Rodriguez-Revenga L, Madrigal I, Pagonabarraga J, Xunclà M, Badenas C, Kulisevsky J, Gomez B, Milà M. Penetrance of FMR1 premutation associated pathologies in fragile X syndrome families. *Eur. J. Hum. Genet.* 2009 Oct;17(10):1359-62.
[65] Hagerman RJ, Leavitt BR, Farzin F, Jacquemont S, Greco CM, Brunberg JA, Tassone F, Hessl D, Harris SW, Zhang L, Jardini T, Gane LW, Ferranti J, Ruiz L, Leehey MA, Grigsby J, Hagerman PJ. Fragile-X-associated tremor/ataxia syndrome (FXTAS) in females with the FMR1 premutation. *Am. J. Hum. Genet.* 2004 May;74(5):1051-6.
[66] Bacalman S, Farzin F, Bourgeois JA, Cogswell J, Goodlin-Jones BL, Gane LW, Grigsby J, Leehey MA, Tassone F, Hagerman RJ. Psychiatric phenotype of the fragile X-associated tremor/ataxia syndrome (FXTAS) in males: newly described fronto-subcortical dementia. *J. Clin. Psychiatry*. 2006 Jan;67(1):87-94.
[67] Bourgeois JA, Cogswell JB, Hessl D, Zhang L, Ono MY, Tassone F, Farzin F, Brunberg JA, Grigsby J, Hagerman RJ. Cognitive, anxiety and mood disorders in the fragile X-associated tremor/ataxia syndrome. *Gen. Hosp. Psychiatry*. 2007 Jul-Aug;29(4):349-56.
[68] Hashimoto R, Backer KC, Tassone F, Hagerman RJ, Rivera SM. An fMRI study of the prefrontal activity during the performance of a working memory task in premutation carriers of the fragile X mental retardation 1 gene with and without fragile X-associated tremor/ataxia syndrome (FXTAS). *J. Psychiatr Res.* 2011 Jan;45(1):36-43.

[69] Shiba M, Bower JH, Maraganore DM, McDonnell SK, Peterson BJ, Ahlskog JE, Schaid DJ, Rocca WA. Anxiety disorders and depressive disorders preceding Parkinson's disease: a case-control study. *Mov. Disord.* 2000 Jul;15(4):669-77.

[70] Duff K, Paulsen JS, Beglinger LJ, Langbehn DR, Stout JC; Predict-HD Investigators of the Huntington Study Group. Psychiatric symptoms in Huntington's disease before diagnosis: the predict-HD study. *Biol. Psychiatry.* 2007 Dec 15;62(12):1341-6. Epub 2007 May 3.

[71] Bourgeois JA, Seritan AL, Casillas EM, Hessl D, Schneider A, Yang Y, Kaur I, Cogswell JB, Nguyen DV, Hagerman RJ. Lifetime prevalence of mood and anxiety disorders in fragile X premutation carriers. *J. Clin. Psychiatry.* 2011 Feb;72(2):175-82. Epub 2010 Aug 24.

[72] Sobesky WE, Hull CE, Hagerman RJ. Symptoms of schizotypal personality disorder in fragile X women. *J. Am. Acad. Child Adolesc. Psychiatry.* 1994 Feb;33(2):247-55.

[73] Cornish K, Kogan C, Turk J, Manly T, James N, Mills A, Dalton A. The emerging fragile X premutation phenotype: evidence from the domain of social cognition. *Brain Cogn.* 2005 Feb;57(1):53-60.

[74] Hocking DR, Kogan CS, Cornish KM. Selective spatial processing deficits in an at-risk subgroup of the fragile X premutation. *Brain Cogn.* 2012 Jun;79(1):39-44. doi: 10.1016/j.bandc.2012.02.005.

[75] Mitchell RJ, Holden JJ, Zhang C et al: FMR1 alleles in Tasmania: a screening study of the special educational needs population. *Clin. Genet.* 2005; 67: 38–46.

[76] Crawford DC, Meadows KL, Newman JL et al: Prevalence and phenotype consequence of FRAXA and FRAXE alleles in a large, ethnically diverse, special education-needs population. *Am. J. Hum. Genet* 1999; 64: 495–507.

[77] Patsalis PC, Sismani C, Hettinger JA et al: Frequencies of ‘grey-zone’ and premutationsize FMR1 CGG-repeat alleles in patients with developmental disability in Cyprus and Canada. *Am. J. Med. Genet* 1999; 84: 195–197.

[78] Mornet E, Chateau C, Simon-Bouy B, Serre JL: The intermediate alleles of the fragile X CGG repeat in patients with mental retardation. *Clin. Genet* 1998; 53: 200–201.

[79] Otsuka S, Sakamoto Y, Siomi H et al: Fragile X carrier screening and FMR1 allele distribution in the Japanese population. *Brain Dev.* 2010; 32: 110–114.

[80] Madrigal I, Xunclà M, Tejada MI, Martínez F, Fernández-Carvajal I, Pérez-Jurado LA, Rodriguez-Revenga L, Milà M. Intermediate FMR1 alleles and cognitive and/or behavioural phenotypes. *Eur. J. Hum. Genet.* 2011 Aug;19(8):921-3.

In: Allelic Forms of the FMR1 Gene
Editor: Montserrat Milà
ISBN: 978-1-63321-914-4

Chapter 5

Clinical Features Associated with *FMR1* Premutation Carriers

M. I. Alvarez-Mora[1,2] ***and L. Rodriguez-Revenga****[1,2,3]
[1]Centre for Biomedical Network Research of Rare Disease (CIBERER), Barcelona, Spain
[2]Biochemistry and Molecular Genetics Department, Hospital Clinic, Barcelona, Spain
[3]IDIBAPS (Institut d'Investigacions Biomèdiques August Pi i Sunyer). Barcelona, Spain

Abstract

The expansion of the CGG trinucleotide located within the 5'UTR of the *FMR1* gene is involved in a growing number of diseases; the most well-established are Fragile X Syndrome (FXS), Fragile X Tremor/Ataxia Syndrome (FXTAS) and Fragile X Primary Ovarian Insufficiency (FXPOI). Whereas full mutation alleles (>200CGGs) are responsible for the FXS, smaller expansions called premutation alleles (55-200CGGs) are associated with FXTAS and FXPOI. Numerous evidence have currently been reported suggesting that premutation alleles give rise to an increased risk for carriers of these alleles in relation to additional medical, psychiatric and cognitive features which occur at a greater frequency than what would be expected for the general population. In this chapter, we review the clinical features including peripheral neuropathy, immune-mediated disorders, migraines and neurocognitive involvement which have been suggested to be associated with premutation alleles. In addition, the current understanding of the pathogenic molecular mechanisms that give rise to the spectrum of *FMR1* premutation associated disorders is also reviewed. Although further research is needed in order to shed light on the factors underlying the common incomplete penetrance applicable to all phenotypes associated with the premutation, it is likely that a combination of environmental and genetic factors with differences in intrinsic susceptibility may modulate the appearance and the severity of these disorders.

Keywords: *FMR1* premutation, fibromyalgia, thyroid disease, peripheral neuropathy

Introduction

In the last years, there has been intense interest in identifying and characterizing the Fragile X premutation-associated phenotypes from the perspective not only of basic science but also of public health given its high prevalence affecting 1:250 females and 1:800 males among the general population [1]. Historically, carriers of *FMR1* premutation (PM) alleles were considered to be clinically unaffected, since the gene is generally not methylated and these individuals do not present intellectual disabilities (ID). The significance of these alleles was generally thought to be their propensity for expansion to the full mutation range (>200 CGG repeats) during maternal transmission resulting in transcriptional silencing of *FMR1* gene, absence of the encoding fragile X mental retardation 1 protein (FMRP) and manifestation of fragile X syndrome (FXS) [2,3]. Although, the features of this expansion are not fully understood, it is currently accepted that they depend on the size of the maternal allele and also on the number of the AGG interruptions within the CGG-repeat track [4]. The AGG interruptions are likely to have stabilizing effects during transmission by decreasing the risk of DNA polymerase slippage during DNA replication [5]. In PM male carriers there is a relative stable transmission, and thus, the risk of expansion to a full mutation is negligible [reviewed in 6]. Despite the belief that PM carriers do not present signs of clinical involvement, prior to the discovery of the gene *FMR1* Cronister and collaborators (1991) [7] reported, higher rates of premature ovarian failure (POF) among women heterozygous for X-chromosome fragility. Later, Allinghan-Hawkins and colleagues (1999) [8] established PM alleles as a significant risk factor for POF based on the study of 760 women in which 16% of PM carriers were affected with POF whereas none of the full mutation carriers and just one (0.4%) of the controls presented with POF [8]. Currently, the association between ovarian deficiency and PM female carriers, namely Fragile X-Primary ovarian Insufficiency (FXPOI), is well-established, presenting an incidence of around 20% in these women and estimated at around 1% among the general population. Chapter 2 describes the features of FXPOI. Ten years after, the first description of a neurodegenerative disorder named Fragile X Tremor Ataxia Syndrome (FXTAS) associated to older adults PM carriers was made by Hagerman and colleagues (2001) [9]. This syndrome is characterized by white matter changes and global brain atrophy, presenting with core features of intention tremor and gait ataxia. Details of FXTAS are described in Chapter 3.

Over the past 10-15 years, an increasingly broad spectrum of clinical manifestations has been related to individuals who are carriers of PM alleles (Table 1). Domains of clinical involvement seen in some, but not all carriers of PM encompass the presence of medical, emotional and cognitive manifestations which have been widely reported to occur more frequently among PM carriers than in the general population. Some of these features have recently been classified as being 'definitely related', 'probably related', 'possibly related' or 'not likely related' to the molecular changes associated with an *FMR1* expansion based on clinical and previously reported data [reviewed in 10].

Coffey and collaborators (2008) [11] reported the first evidence of an expanded clinical phenotype of women with the PM. In this study, PM female carriers without core features of FXTAS showed significantly more complaints of chronic muscle pain, persistent paraesthesias in the extremities, and a history of tremor than controls. Furthermore, a significantly greater presence of medical co-morbidity was detected in females with definite or probable FXTAS, with an increased prevalence of thyroid disease, hypertension, seizures, peripheral neuropathy, fibromyalgia compared with controls [11]. In addition, some of the comorbidites associated with FXTAS beyond central nervous system involvement, specifically peripheral neuropathy [11,12] and neuroendocrine dysfunction [11,13-15] have also been associated with PM carriers without FXTAS. Moreover, there is increasing evidence that young PM carriers present increased rates of neurodevelopmental phenotypes such as attention-deficit hyperactivity disorder (ADHD), autism spectrum disorder (ASD) and seizures [reviewed in 16]. Furthermore, increased rates of psychiatric involvement, particularly depression and anxiety have also been associated with FXTAS among adult

PM carriers [reviewed in 17]. Neuropsychiatric aspects are discussed in Chapter 4.

Table 1.Clinical manifestations associated with some *FMR1* premutation carriers

	Cohort studied*	References
Immune mediated disorders		
Fibromyalgia	PM females	[11-15]
Thyroid disease	PM females	[11,13,15]
Irritable bowel syndrome	PM females	[15]
Neurodevelopmental Phenotypes		
Working memory deficiencies	PM males	[27,28]
Language dysfluencies	PM females	[29]
Spatiotemporal processing impairment	Young-adult PM carriers	[32]
Arithmetic weaknesses	PM females	[30,31]
Developmental Delay	PM carriers	[75]
Reproductive features		
Ovarian Insufficiency (FXPOI)	PM females	[reviewed in 10]
Obstetric and perinatal difficulties	PM females	
Estrogen-deficiency related conditions	FXPOI PM females	
Autonomic dysfunction		
Impotence	FXTAS PM males	[56]
Hypertension	FXTAS PM both	[11,67]
Bowel and bladder incontinence	FXTAS PM both	[18,68]
Neurocognitive and phychiatric Involvement		
Depression	PM females	[69-71]
Anxiety disorders	PM females	
Mood disorders	PM females	
Seizures	PM male children	[72]
ASD	PM male children	[72,73]
ADHD	PM females	[74]
Other clinical manifestations		
Migraine	PM carriers both	[25]
Peripheral neuropathy	PM carriers both	[11,76]

Clinical Features Associated with *FMR1* Premutation Carriers

Neuropathy

Peripheral neuropathy is characterized by a damaging of peripheral nerves which carry information to and from the brain as well as to and from the spinal cord to the rest of the body. Depending on the type of nerve affected it may promote impaired sensation, movement, gland or organ function. The association between neuropathy and *FMR1* alleles was first reported in PM carriers with FXTAS [9,18]. Thereafter, Berry-Kravis and collaborators (2007) [12] reported the first evidence that signs of neuropathy on clinical examination are associated with PM carrier status based on neurological examination data from 207 unrelated individuals. Results from this study revealed that the degree of clinical involvement strongly correlated with the CGG repeat length in males since these individuals presented significantly higher mean scores in the neuropathy screening scale score (P=0.0014), the vibration score (P=0.0015), and the reflex score (P=0.0014) than sex-matched controls suggesting that PM male carriers present higher impairment of both distal vibratory sense and reflexes [12]. The lack of significant differences among PM female carriers presumably reflects the broad variation in clinical involvement among carriers as a result of variation in the X-chromosome activation ratio as well as the decreased penetrance of the clinical manifestation due to the protective effect of the second non-mutated X chromosome [12]. Afterwards, Coffey and collaborators (2008) [11] demonstrated that females with PM alleles also presented significantly higher signs of neuropathy based on findings from 128 PM female carriers without FXTAS. In this study it was shown that these women presented a significant rate of numbness and tingling and muscle pain in the extremities, albeit evidence of neuropathy is increased in PM female carriers presenting with FXTAS [11]. It has been suggested that neuropathic symptoms in PM female carriers are manifested together with the emergence of FXTAS disease [reviewed in 10].

Immune-Mediated Disorders

Numerous reports support elevated rates of immune-mediated disorders (IMD) in PM female carriers, particularly regarding hypothyroidism and fibromyalgia [11, 13-15]. In contrast, IMDs have not been evidenced among males with the premutation, likely due to the relative rarity of these disorders among the general male population. In a recent study, Winarni and colleagues (2012) [15] examined the relative likelihood of large clinical manifestations among 344 female carriers of PM alleles and 72 controls including autoimmune thyroid disorder, multiple sclerosis, Sjögren syndrome, rheumatoid arthritis, systemic lupus erythematosus, Raynaud's phenomenon, irritable bowel syndrome and optic neuritis. The results of this study evidenced that among women over 40 years of age 46.54% of PM females without FXTAS experienced one or more of the IMDs surveyed, and the prevalence increased to about 72.73% for those with FXTAS compared to 31.58% for the control group [15]. With respect to FXPOI, both groups of PM females carriers present higher odds ratios of IMDs compared to controls, and similarly, when considering FXTAS symptoms, the odds ratio of IMDs among PM female carriers presenting with FXPOI is about 2.4-fold higher when compared to those without FXPOI. Moreover, these authors found that an autoimmune thyroid disorder was the most common IMD followed by fibromyalgia and irritable bowel syndrome [15].

Increased penetrance for both thyroid disease and fibromyalgia has been broadly reported among PM female carriers [13], although the penetrance of these disorders is highly variable among the general population since it increases with age. For thyroid disease, it has been suggested that the association with PM alleles may be more relevant in older women [10] due to the lack of statistical significance when considering women between 18 and 50 years of age [19]. Nonetheless, in the general population the penetrance estimated for thyroid disorders is of around 10% [20] and around 2-4% for fibromyalgia whereas it has been estimated to be around 15.9% and 24.4%, respectively, among PM female carriers [13]. Conversely, in 700 unrelated Spanish patients with fibromyalgia the frequency of PM alleles did not significantly differ from the estimated rate in the general population [21]. In contrast, another study found a higher incidence of PM alleles among a Spanish female fibromyalgia cohort. Indeed, the incidence of PM alleles was 1 of 88, being 1 of 250 in females in the general population [22]. These controversial results are likely to be caused by a sample size effect since the data reported by Martorell and colleagues (2012) [22] were based on the screening of 353 females whereas the data presented by Rodriguez-Revenga and coworkers (2013) [21] were based on 700 samples. Nonetheless, it has been shown that the pathophysiology of fibromyalgia involves hyperexcitability of central neurons through several synaptic and neurotransmitter/neurochemical mechanisms [reviewed in 23] suggesting that it could arise through an alteration of pain neurotransmissor mechanisms among PM female carriers [14].

Finally, it has been recently demonstrated that individual carriers of PM alleles present an immune dysregulation and decreased immune responses when compared with healthy controls [24]. Moreover, it has been found that PM carriers present a reduction in the levels of cytokine production which is negatively associated with CGG repeat length, mainly with IL-12 production [24]. Furthermore, these women have also been shown to present a decrease in the relative levels of the surface marker CD25 in T cells suggesting potential differences in the activation of T-cells that regulate immune response [24].

Migraines

Au and collaborators (2013) [25] have recently reported that PM carriers show increased rates in the prevalence of migraines based on physical and medical examination of 315 PM carriers (203 females and 112 males) and 154 controls (83 females and 71 males). Migraine is a neurologic disorder characterized by light and sound sensitivity and pulsatile pain, which is thought to have a polygenic and mutlifactorial etiology. The prevalence of migraine among the general population is estimated around 27.3% in women and 9.7% in men [26], reaching up to 54.2% and 26.79% among female and male PM carriers, respectively, both resulting in statistical significant differences [25]. In addition, this significance was obtained considering both those affected with and also those without FXTAS, adjusted for age. However, the risk of migraine headaches was not correlated with either CGG repeats or *FMR1* mRNA expression [25].

Neurocognitive Features

The expanded range of clinical involvement associated with PM carriers also includes an alteration of various cognitive domains including executive function, working memory and arithmetic skills which become apparent even in young individuals, with a usually more progressive course in PM individuals than in the general population [reviewed in 10]. However, neurocognitive deficits are reportedly more frequent in male than in female carriers of PM alleles. Results reported by Kogan and collaborators (2008) [27] revealed that the CGG expansion confers a significant risk for working memory difficulties based on a controlled study of 40 PM male carriers without manifest symptoms of FXTAS. In addition, Cornish and colleagues (2009) [28] reported neuropsychological measures in PM males regarding core subcomponents of working memory such as verbal memory, visual–spatial memory, and central executive memory revealing that PM males present specific vulnerability in executive control of memory including tasks requiring simultaneous manipulation and storage of new information, regardless of the presence of FXTAS symptoms. These authors revealed an impairment of the central executive working memory among PM male carriers without FXTAS, which was significantly correlated with larger CGG repeat expansions, whereas FXTAS patients demonstrated a more general impairment in terms of phonological working memory in addition to central executive working memory [28]. Moreover, language dysfluencies associated with deficits in organization and planning have been evidenced among PM female carriers [29]. Past research has demonstrated that language dysfluencies are an indicator of executive functioning deficits which are characteristic of other neurodegenerative disorders such as Parkinson and Alzheimer diseases.

Regarding arithmetic skills it has been suggested that PM female carriers show weaknesses in mathematical tasks [30] and this has recently been supported by other groups [31]. Furthermore, it has been demonstrated that PM carriers from 19 to 45 years of age show impairment in spatiotemporal processing which may underlie the impairments observed in arithmetic skills among these individuals since the representations of space and time provide the foundation for an understanding of numbers [32].

Although further research is needed, it has been suggested that determining whether cognitive impairments are detectable in PM carriers without FXTAS should be prudent since it may not only be an early indicator of cognitive decline in PM carriers [29] but could also be used as a biomarker of disease progression if these features precede motor impairment [32].

Current Understanding of the Molecular Mechanisms Underlying Premutation-Associated Pathologies

Premutation-associated phenotypes have been mainly attributed to a pathogenic mechanism involving a gain-of-function toxicity of the expanded *FMR1* mRNA, a process entirely distinct from the FMRP deficiency responsible for the FXS phenotype [reviewed in 33]. This observation was based on the restriction of these clinical phenotypes to the premutation range in which the molecular signature is a 2-8 fold increase in the expression of the PM mRNA and, paradoxically, a slight reduction in Fragile X Mental Retardation Protein (FMRP) levels [34]. The mechanisms underlying the increased transcriptional activity of the PM alleles remain unclear, however, it has been reported that these alleles use differential transcriptional start sites leading to different expression compared to non-expanded *FMR1* alleles [35]. On other hand, it has also been suggested that the reduction of FMRP in PM carriers may promote an added contribution to the clinical involvement observed in both children and adults related to phenotypes associated with reduced cognition and disturbed behavior [36]. However, the FMRP deficiency cannot be a driving factor in the PM-associated disorders since FXTAS and FXPOI are not experienced by full mutation carriers. Despite wide reports that the levels of FMRP expression are only slightly decreased in patients with FXTAS, most of these measurements have been performed using lymphocytes or whole blood samples rather than brain tissue, in which changes are likely to be more robust [reviewed in 37]. The efficiency of FMRP translation is associated with the length of the CGG expansion in light of impairment in translation of larger CGG tracks by interference with ribosomal scanning through the 5'UTR, thereby preventing appropriate loading of the expanded *FMR1* mRNA into polyribosomal complexes [38-40].

It is currently considered that the dual mechanism of involvement, the excess of PM mRNA expression and the decrease in FMRP translation, is a double hit which may promote phenotypic features of FXS and PM associated disorders [reviewed in 33, 41]. Notwithstanding, the reduction of FMRP synthesis, the phenotype of PM carriers, is different from carriers of full mutation alleles since milder protein deficiency among PM carriers usually leads to mild developmental problems with these individuals having higher IQs and less severe behavioral problems than those with FXS [reviewed in 42]. A FMRP deficit has been correlated with lowered activity of the amygdala among PM male carriers compared to controls on functional magnetic resonance imaging whereas increased levels of expanded mRNA have been strongly associated with obsessive–compulsive symptoms and psychoticism in PM male carriers [43,44]. It has also been described that RNA toxicity leads to the up-regulation of the heat shock proteins Hsp70 and αB-crystallin, which may stimulate immune dysregulation [15]. Regarding migraines, their association with mitochondrial dysfunction is well established. Interestingly, there is evidence pointing to a deregulation of mitochondrial function among PM carriers [45-47], therefore the increased prevalence of migraines in PM carriers may be the result of RNA toxicity leading to mitochondrial deregulation [25]. Likewise, RNA toxicity has also been proposed to shed light on the high rate of thyroid dysfunction among females with the PM by causing a direct effect on the hypothalamic-pituitary-adrenal axis or on the thyroid gland. Additionally, RNA toxicity has also been proposed to promote an autoimmune mechanism or apoptosis in thyroid cells [11]. However, data reported by Cunningham and coworkers (2011) [48] show that the presence of PM alleles promotes migration defects in the neocortex and altered expression of neuronal lineage markers among embryonic PM mice. These results support the hypothesis that the role of the RNA toxicity may be restricted to the initial triggering events since many features of the neuronal and astrocytic cellular phenotype observed in FXTAS patients are already present in the neonatal period, suggesting that the clinical involvement among children carriers of PM alleles may be manifestations of this early, non-degenerative process [reviewed in 33].

The sequestration hypothesis of RNA toxicity was first proposed and established for myotonic dystrophy type 1 caused by an expansion of a CTG repeat in the 3'UTR of *DMPK* gene [reviewed in 49]. Particularly, in Fragile X PM the model hypothesized that expanded CGG repeats form hairpin loops which are sticky and recruit an excess of specific RNA-binding proteins, resulting in a functional insufficiency of the sequestered proteins and leading to cell dysfunction and death [50]. Although the initial triggering events in these disorders are based on RNA toxicity, this model does not provide evidence regarding cell sickening and death. In this way, there are several candidate downstream pathways, although alterations of both mitochondrial function and calcium regulation are emerging as core mediators of cellular deregulation and dysfunction [reviewed in 31, 39] The increased expression of the expanded *FMR1* mRNA is thought to be the main cause of clinical involvement in PM carriers since the *FMR1* mRNA and the sequestered proteins form aggregates leading to intranuclear inclusions present in several tissues. Interestingly, the expanded mRNA of *FMR1* is detected within the inclusion whereas FMRP is not present [51]. Furthermore, proteomic analysis of these inclusions revealed a large number of proteins presented in the aggregates, including the RNA binding protein hnRNP A2/B1, the nuclear envelope protein lamin A/C, the small heat shock protein αB-crystallin [52], the splicing factor Sam68 [53] and part of the microRNA processor complex DGCR8 [50]. Indeed, it has recently been demonstrated that the sequestration of DGCR8 promotes the deregulation of microRNAs biogenesis, suggesting a central role of DGCR8 as an inductor of RNA toxicity by leading to cell dysfunction and cell loss [50].

Intranuclear inclusions, which represent the neuropathological hallmark of FXTAS [52], have been also detected among PM carriers through different cell types including the central and peripheral nervous system and other tissue including the adrenal glands, the testes, pancreas and heart [54-58]. Recently,

Hunsaker and colleagues (2011) [58] reported autopsy findings from ten PM carriers with FXTAS in which intranuclear inclusions were detected throughout multiple tissues including the hypothalamic-pituitary-adrenal axis, pineal gland, cardiac conduction system, peripheral nerves and autonomic ganglia, the thyroid gland, the digestive system, the testes and pancreas. The broad distribution of these inclusions suggests that many organ systems may be affected by RNA toxicity. Nevertheless, it is necessary to study the processes underlying inclusion formation in depth to address whether they themselves are toxic or reflect cellular dysfunction [58].

Furthermore, additional mechanisms have been proposed as possible triggering events in the PM-associated disorders, although most evidence support CGG-repeat mediated protein sequestration [reviewed in 33,37]. These mechanisms include a RNA-mediated protein aggregation model whereby the CGGs contained within the *FMR1* mRNA might promote a conformational transition in proteins with prion-like domains that may initiate a cascade of protein aggregation similar to what occurs in amyloid plaque formation in Alzheimer's disease. Moreover, the production of a toxic polyglycine peptide has also been proposed as an alternative toxicity model by a non-AUG-initiated (RAN) translation [59]. In addition, a specific splicing isoform is detected exclusively with transcripts of PM alleles suggesting a possible role of antisense transcripts generated at the *FMR1* locus [60].

Otherwise very little attention has been focused on the other end of the spectrum, the so-called "low-normal" numbers of CGG repeats, up to 23 trinucleotide repeats. In this framework, data based on genotype–phenotype correlations have recently been reported suggesting that this range of CGG repeats may have substantial implications for cognitive functioning, cancer, and the odds of having children with neurodevelopmental or neuropsychiatric conditions [61]. Despite these range of CGG number not being associated with altered FMRP synthesis, Chen and co-workers (2003) [62] reported that the efficiency of FMRP translation was based on the number of CGG repeats, conferring the greatest efficiency of protein synthesis to the allele of 30 repeats. Thus, inefficient translation may be related to the clinical manifestations associated with the low numbers of CGG repeats. Likewise, Ramocki and Zoghbi (2008) [63] suggested that imbalances in homeostatic controls of multiple genes including *FMR1* may partially promote the appearance of neurodevelopmental and neuropsychiatric disorders.

At this point, the difficulty in understanding PM-associated pathologies lies in the inability of prediction of which PM carriers will develop any of these phenotypes. The incomplete penetrance across the phenotypic spectrum is likely to be associated with a combination of genetic and environmental factors which may confer specific vulnerability to PM carriers (Figure 1).

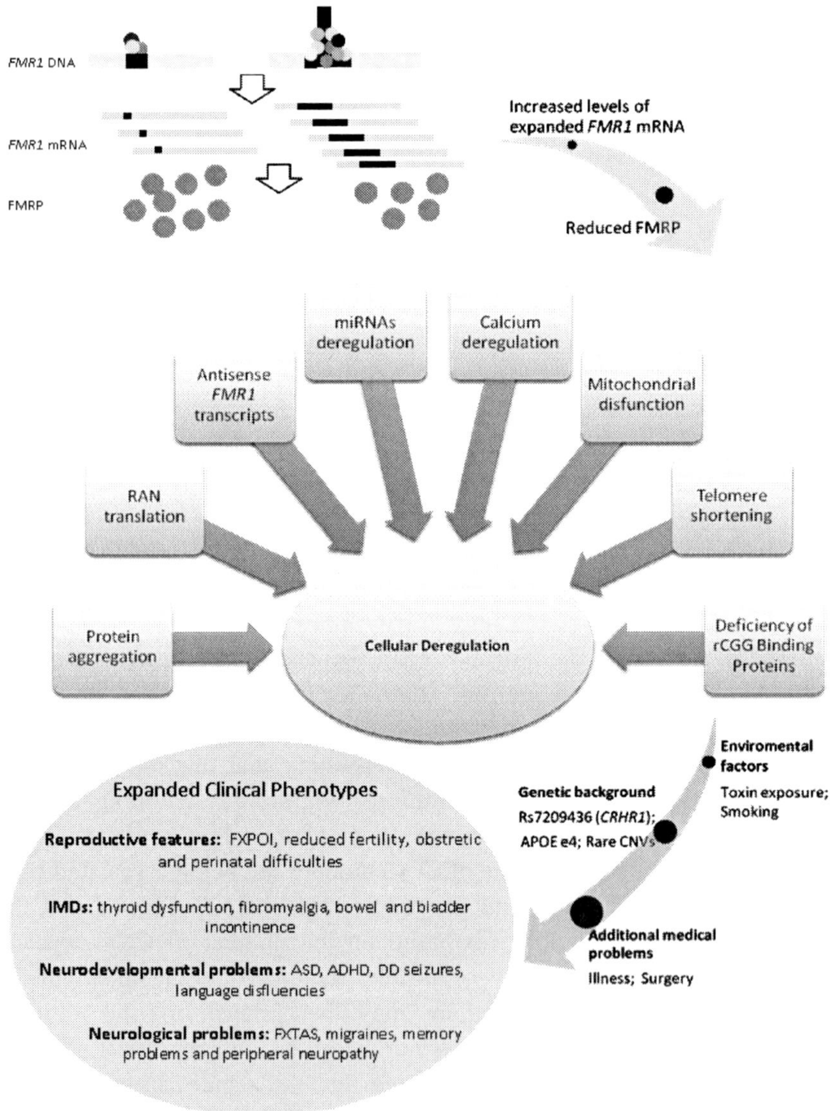

Figure 1. Diagram of the potential players contributing to the clinical involvement associated to PM carriers.

Genetic factors that may contribute to the PM-associated disorders include CGG repeat length, expression levels of the expanded *FMR1* mRNA, aberrant translation of the repeat sequence as well as genomic changes in other regions of the genome. Within this framework, specific polymorphisms of the *CRHR1* gene have been associated with female clinical involvement (rs7209436), particularly with depression and anxiety mainly as this gene regulates the expression and release of ACTH from the anterior pituitary gland which, in turn, stimulates the release of cortisol from the adrenal cortex [64]. Furthermore, risk factors for other neurodegenerative disorders such as allele ε4 of the *APOE* gene may also influence the risk of FXTAS as a higher frequency of these alleles has been reported in PM carriers with compared to those without FXTAS [65]. Moreover, it has recently been suggested that individuals who are carriers of PM alleles presenting with ID, seizures or ASD are likely to have a second hit since PM carriers show a significant enrichment (P=2.27e-07) of CNVs compared to controls [66]. Furthermore, these authors found an association between the presence of rare CNVs (not detected in 8000 controls) among PM carriers with either autistic traits or neurological involvement, suggesting that they may have a possible role in this phenotypic variability [66]. Regarding environmental factors, it has been suggested that smoking, prolonged surgery with anesthesia, drug and alcohol abuse or the stress of carrying a child with FXS are also puzzling factors which may act as additional determinants for the phenotypic variability among PM individuals [reviewed in 41,42]. Finally, further longitudinal studies are required to determine the context in which any of the PM-associated phenotypes are developed and what protective factors might reduce the risks of more negative outcomes [reviewed in 10].

Conclusion

Overall, it is currently accepted that PM alleles led to multiple distinct clinical features which are present at a greater frequency among PM carriers than what would be expected in the general population. However, the association of the PM with the phenotypes reviewed in this chapter is less well established than its association with FXTAS and FXPOI. Although further research is needed in order to shed light on the factors underlying the common incomplete penetrance applicable to all phenotypes associated with the PM, a combination of environmental and genetic factors with differences in intrinsic susceptibility likely modulate the appearance and the severity of these disorders.

References

[1] Hagerman PJ. The fragile X prevalence paradox. J Med. Genet.. 2008 Aug;45(8):498-9. Erratum in: *J. Med. Genet.* 2008 Nov;45(11):768.

[2] Rousseau F, Heitz D, Biancalana V, Blumenfeld S, Kretz C, Boué J, Tommerup N, Van Der Hagen C, DeLozier-Blanchet C, Croquette MF, et al. Direct diagnosis by DNA analysis of the fragile X syndrome of mental retardation. *N. Engl. J. Med.* 1991 Dec 12;325(24):1673-81.

[3] Nolin SL, Lewis FA 3rd, Ye LL, Houck GE Jr, Glicksman AE, Limprasert P, Li SY, Zhong N, Ashley AE, Feingold E, Sherman SL, Brown WT. Familial transmission of the FMR1 CGG repeat. *Am. J. Hum. Genet.* 1996 Dec;59(6):1252-61.

[4] Yrigollen CM, Tassone F, Durbin-Johnson B, Tassone F. The role of AGG interruptions in the transcription of PM alleles. *PLoS One.* 2011;6(7):e21728.

[5] Gacy AM, Goellner G, Juranić N, Macura S, McMurray CT. Trinucleotide repeats that expand in human disease form hairpin structures in vitro. *Cell.* 1995 May 19;81(4):533-40.

[6] Brouwer JR, Willemsen R, Oostra BA. Microsatellite repeat instability and neurological disease. *Bioessays.* 2009 Jan;31(1):71-83. doi: 10.1002/bies.080122.

[7] Cronister A, Schreiner R, Wittenberger M, Amiri K, Harris K, Hagerman RJ. Heterozygous fragile X female: historical, physical, cognitive, and cytogenetic features. *Am. J. Med. Genet.* 1991 Feb-Mar;38(2-3):269-74.

[8] Allingham-Hawkins DJ, Babul-Hirji R, Chitayat D, Holden JJ, Yang KT, Lee C, Hudson R, Gorwill H, Nolin SL, Glicksman A, Jenkins EC, Brown WT, Howard-Peebles PN, Becchi C, Cummings E, Fallon L, Seitz S, Black SH, Vianna-Morgante AM, Costa SS, Otto PA, Mingroni-Netto RC, Murray A, Webb J, Vieri F, et al. Fragile X premutation is a significant risk factor for premature ovarian failure: the International Collaborative POF in Fragile X study--preliminary data. *Am. J. Med. Genet.* 1999 Apr 2;83(4):322-5.

[9] Hagerman RJ, Leehey M, Heinrichs W, Tassone F, Wilson R, Hills J, Grigsby J, Gage B, Hagerman PJ. Intention tremor, parkinsonism, and

generalized brain atrophy in male carriers of fragile X. *Neurology.* 2001 Jul 10;57(1):127-30.

[10] Wheeler AC, Bailey DB Jr, Berry-Kravis E, Greenberg J, Losh M, Mailick M, Milà M, Olichney JM, Rodriguez-Revenga L, Sherman S, Smith L, Summers S, Yang JC, Hagerman R. Associated features in females with an PM. *J Neurodev Disord.* 2014;6(1):30. Review.

[11] Coffey SM, Cook K, Tartaglia N, Tassone F, Nguyen DV, Pan R, Bronsky HE, Yuhas J, Borodyanskaya M, Grigsby J, Doerflinger M, Hagerman PJ, Hagerman RJ. Expanded clinical phenotype of women with the PM. *Am. J. Med. Genet. A.* 2008 Apr 15;146A(8):1009-16.

[12] Berry-Kravis E, Goetz CG, Leehey MA, Hagerman RJ, Zhang L, Li L, Nguyen D, Hall DA, Tartaglia N, Cogswell J, Tassone F, Hagerman PJ. Neuropathic features in fragile X premutation carriers. *Am. J. Med. Gene.t A.* 2007 Jan 1;143A(1):19-26.

[13] Rodriguez-Revenga L, Madrigal I, Pagonabarraga J, Xunclà M, Badenas C, Kulisevsky J, Gomez B, Milà M. Penetrance of FMR1 premutation associated pathologies in fragile X syndrome families. *Eur. J. Hum. Genet.* 2009 Oct;17(10):1359-62.

[14] Leehey MA, Legg W, Tassone F, Hagerman R. Fibromyalgia in fragile X mental retardation 1 gene premutation carriers. *Rheumatology* (Oxford). 2011 Dec;50(12):2233-6.

[15] Winarni TI, Chonchaiya W, Sumekar TA, Ashwood P, Morales GM, Tassone F, Nguyen DV, Faradz SM, Van de Water J, Cook K, Hamlin A, Mu Y, Hagerman PJ, Hagerman RJ. Immune-mediated disorders among women carriers of fragile X premutation alleles. *Am. J. Med. Genet. A.* 2012 Oct;158A(10):2473-81.

[16] Besterman AD, Wilke SA, Mulligan TE, Allison SC, Hagerman R, Seritan AL, Bourgeois JA. Towards an Understanding of Neuropsychiatric Manifestations in Fragile X Premutation Carriers. *Future Neurol.* 2014 Mar;9(2):227-239.

[17] Hunter JE, Abramowitz A, Rusin M, Sherman SL. Is there evidence for neuropsychological and neurobehavioral phenotypes among adults without FXTAS who carry the PM? A review of current literature. *Genet Med.* 2009 Feb;11(2):79-89.

[18] Jacquemont S, Hagerman RJ, Leehey M, Grigsby J, Zhang L, Brunberg JA, Greco C, Des Portes V, Jardini T, Levine R, Berry-Kravis E, Brown WT, Schaeffer S, Kissel J, Tassone F, Hagerman PJ. Fragile X premutation tremor/ataxia syndrome: molecular, clinical, and neuroimaging correlates. *Am. J. Hum. Genet.* 2003 Apr;72(4):869-78.

[19] Hunter JE, Rohr JK, Sherman SL. Co-occurring diagnoses among FMR1 premutation allele carriers. *Clin. Genet.* 2010 Apr;77(4):374-81.

[20] Canaris GJ, Manowitz NR, Mayor G, Ridgway EC. The Colorado thyroid disease prevalence study. *Arch. Intern Med.* 2000 Feb 28;160(4):526-34.

[21] Rodriguez-Revenga L, Madrigal I, Blanch-Rubió J, Elurbe DM, Docampo E, Collado A, Vidal J, Carbonell J, Estivill X, Mila M. Screening for the presence of PM alleles in women with fibromyalgia. *Gene.* 2013 Jan 10;512(2):305-8.

[22] Martorell L, Tondo M, Garcia-Fructuoso F, Naudo M, Alegre C, Gamez J, Genovés J, Poo P. Screening for the presence of FMR1 premutation alleles in a Spanish population with fibromyalgia. *Clin. Rheumatol.* 2012 Nov;31(11):1611-5.

[23] Yunus MB. Role of central sensitization in symptoms beyond muscle pain, and the evaluation of a patient with widespread pain. *Best Pract. Res. Clin. Rheumatol.* 2007 Jun;21(3):481-97. Review.

[24] Careaga M, Rose D, Tassone F, Berman RF, Hagerman R, Ashwood P. Immune dysregulation as a cause of autoinflammation in fragile X premutation carriers: link between FMRI CGG repeat number and decreased cytokine responses. *PLoS One.* 2014 Apr 9;9(4):e94475.

[25] Au J, Akins RS, Berkowitz-Sutherland L, Tang HT, Chen Y, Boyd A, Tassone F, Nguyen DV, Hagerman R. Prevalence and risk of migraine headaches in adult fragile X premutation carriers. *Clin Genet.* 2013 Dec;84(6):546-51.

[26] Lipton RB, Stewart WF, Diamond S, Diamond ML, Reed M. Prevalence and burden of migraine in the United States: data from the American Migraine Study II. *Headache* 2001: 41 (7): 646–657.

[27] Kogan CS, Turk J, Hagerman RJ, Cornish KM. Impact of the Fragile X mental retardation 1 (FMR1) gene premutation on neuropsychiatric functioning in adult males without fragile X-associated Tremor/Ataxia syndrome: a controlled study. *Am. J. Med. Genet. B Neuropsychiatr Genet.* 2008 Sep 5;147B(6):859-72.

[28] Cornish KM, Kogan CS, Li L, Turk J, Jacquemont S, Hagerman RJ. Lifespan changes in working memory in fragile X premutation males. *Brain Cogn.* 2009 Apr;69(3):551-8.

[29] Sterling AM, Mailick M, Greenberg J, Warren SF, Brady N. Language dysfluencies in females with the PM. *Brain Cogn.* 2013 Jun;82(1):84-9.

[30] Lachiewicz AM, Dawson DV, Spiridigliozzi GA, McConkie-Rosell A. Arithmetic difficulties in females with the fragile X premutation. *Am. J. Med. Genet. A.* 2006 Apr 1;140(7):665-72.

[31] Semenza C, Bonollo S, Polli R, Busana C, Pignatti R, Iuculano T, Maria Laverda A, Priftis K, Murgia A. Genetics and mathematics: FMR1 premutation female carriers. *Neuropsychologia*. 2012 Dec;50(14):3757-63

[32] Wong LM, Goodrich-Hunsaker NJ, McLennan Y, Tassone F, Harvey D, Rivera SM, Simon TJ. Young adult male carriers of the fragile X premutation exhibit genetically modulated impairments in visuospatial tasks controlled for psychomotor speed. *J. Neurodev Disord.* 2012 Nov 13;4(1):26.

[33] Hagerman P. Fragile X-associated tremor/ataxia syndrome (FXTAS): pathology and mechanisms. *Acta Neuropathol.* 2013 Jul;126(1):1-19.

[34] Tassone F, Beilina A, Carosi C, Albertosi S, Bagni C, Li L, Glover K, Bentley D, Hagerman PJ. Elevated FMR1 mRNA in premutation carriers is due to increased transcription. *RNA*. 2007 Apr;13(4):555-62.

[35] Tassone F, De Rubeis S, Carosi C, La Fata G, Serpa G, Raske C, Willemsen R, Hagerman PJ, Bagni C. Differential usage of transcriptional start sites and polyadenylation sites in FMR1 premutation alleles. *Nucleic Acids Res*. 2011 Aug;39(14):6172-85.

[36] Tassone F, Hagerman RJ, Taylor AK, Mills JB, Harris SW, Gane LW, Hagerman PJ (2000) Clinical involvement and protein expression in individuals with the FMR1 premutation. *Am. J. Med. Genet.* 91(2):144–152

[37] Renoux AJ, Todd PK. Neurodegeneration the RNA way. *Prog. Neurobiol.* 2012 May;97(2):173-89.

[38] Primerano B, Tassone F, Hagerman RJ, Hagerman P, Amaldi F, Bagni C. Reduced FMR1 mRNA translation efficiency in fragile X patients with premutations. *RNA*. 2002 Dec;8(12):1482-8.

[39] Ludwig AL, Raske C, Tassone F, Garcia-Arocena D, Hershey JW, Hagerman PJ. Translation of the FMR1 mRNA is not influenced by AGG interruptions. *Nucleic Acids Res.* 2009 Nov;37(20):6896-904.

[40] Ludwig AL, Hershey JW, Hagerman PJ. Initiation of translation of the FMR1 mRNA Occurs predominantly through 5'-end-dependent ribosomal scanning. *J. Mol. Biol.* 2011 Mar 18;407(1):21-34.

[41] Hagerman R, Hagerman P. Advances in clinical and molecular understanding of the FMR1 premutation and fragile X-associated tremor/ataxia syndrome. *Lancet Neurol.* 2013 Aug;12(8):786-98.

[42] Polussa J, Schneider A, Hagerman R. (2014) Molecular Advances Leading to Treatment Implications for Fragile X Premutation Carriers. *Brain Disord Ther* 3:2.

[43] Hessl D, Tassone F, Loesch DZ, Berry-Kravis E, Leehey MA, Gane LW, Barbato I, Rice C, Gould E, Hall DA, Grigsby J, Wegelin JA, Harris S, Lewin F, Weinberg D, Hagerman PJ, Hagerman RJ. Abnormal elevation of FMR1 mRNA is associated with psychological symptoms in individuals with the fragile X premutation. *Am. J. Med. Genet. B Neuropsychiatr Genet.* 2005; 139B (1):115–121.

[44] Hoem G, Raske CR, Garcia-Arocena D, Tassone F, Sanchez E, Ludwig AL, Iwahashi CK, Kumar M, Yang JE, Hagerman PJ. CGG-repeat length threshold for FMR1 RNA pathogenesis in a cellular model for FXTAS. *Hum. Mol. Genet.* 2011; 20 (11):2161–2170.

[45] Ross-Inta C, Omanska-Klusek A, Wong S, Barrow C, Garcia-Arocena D, Iwahashi C, Berry-Kravis E, Hagerman RJ, Hagerman PJ, Giulivi C. Evidence of mitochondrial dysfunction in fragile X-associated tremor/ataxia syndrome. *Biochem. J.* 2010 Aug 1;429(3):545-52.

[46] Napoli E, Ross-Inta C, Wong S, Omanska-Klusek A, Barrow C, Iwahashi C, Garcia-Arocena D, Sakaguchi D, Berry-Kravis E, Hagerman R, Hagerman PJ, Giulivi C. Altered zinc transport disrupts mitochondrial protein processing/import in fragile X-associated tremor/ataxia syndrome. *Hum. Mol. Genet.* 2011 Aug 1;20(15):3079-92.

[47] Mateu-Huertas E, Rodriguez-Revenga L, Alvarez-Mora MI, Madrigal I, Willemsen R, Milà M, Martí E, Estivill X. Blood expression profiles of fragile X premutation carriers identify candidate genes involved in neurodegenerative and infertility phenotypes. *Neurobiol. Dis.* 2014 May;65:43-54.

[48] Cunningham CL, Martínez Cerdeño V, Navarro Porras E, Prakash AN, Angelastro JM, Willemsen R, Hagerman PJ, Pessah IN, Berman RF, Noctor SC. Premutation CGG-repeat expansion of the Fmr1 gene impairs mouse neocortical development. *Hum. Mol. Genet.* 2011 Jan 1;20(1):64-79.

[49] Todd PK, Paulson HL. RNA-mediated neurodegeneration in repeat expansion disorders. *Ann. Neurol.* 2010 Mar;67(3):291-300. Review.

[50] Sellier C, Freyermuth F, Tabet R, Tran T, He F, Ruffenach F, Alunni V, Moine H, Thibault C, Page A, Tassone F, Willemsen R, Disney MD, Hagerman PJ, Todd PK, Charlet-Berguerand N. Sequestration of DROSHA and DGCR8 by expanded CGG RNA repeats alters

microRNA processing in fragile X-associated tremor/ataxia syndrome. *Cell Rep.* 2013 Mar 28;3(3):869-80.

[51] Tassone F, Iwahashi C, Hagerman PJ. FMR1 RNA within the intranuclear inclusions of fragile X-associated tremor/ataxia syndrome (FXTAS). *RNA Biol.* 2004 Jul;1(2):103-5.

[52] Iwahashi CK, Yasui DH, An HJ, Greco CM, Tassone F, Nannen K, Babineau B, Lebrilla CB, Hagerman RJ, Hagerman PJ. Protein composition of the intranuclear inclusions of FXTAS. *Brain.* 2006 Jan;129(Pt 1):256-71.

[53] Sellier C, Rau F, Liu Y, Tassone F, Hukema RK, Gattoni R, Schneider A, Richard S, Willemsen R, Elliott DJ, Hagerman PJ, Charlet-Berguerand N. Sam68 sequestration and partial loss of function are associated with splicing alterations in FXTAS patients. *EMBO J.* 2010 Apr 7;29(7):1248-61.

[54] Louis E, Moskowitz C, Friez M, Amaya M, Vonsattel JP. Parkinsonism, dysautonomia, and intranuclear inclusions in a fragile X carrier: a clinical-pathological study. *Mov. Disord.* 2006 Mar;21(3):420-5.

[55] Greco CM, Berman RF, Martin RM, Tassone F, Schwartz PH, Chang A, Trapp BD, Iwahashi C, Brunberg J, Grigsby J, Hessl D, Becker EJ, Papazian J, Leehey MA, Hagerman RJ, Hagerman PJ. Neuropathology of fragile X-associated tremor/ataxia syndrome (FXTAS). *Brain.* 2006 Jan;129(Pt 1):243-55.

[56] Greco CM, Soontrapornchai K, Wirojanan J, Gould JE, Hagerman PJ, Hagerman RJ. Testicular and pituitary inclusion formation in fragile X associated tremor/ataxia syndrome. *J. Urol.* 2007 Apr;177(4):1434-7.

[57] Gokden M, Al-Hinti JT, Harik SI. Peripheral nervous system pathology in fragile X tremor/ataxia syndrome (FXTAS). *Neuropathology.* 2009 Jun;29(3):280-4.

[58] Hunsaker MR, Greco CM, Spath MA, Smits AP, Navarro CS, Tassone F, Kros JM, Severijnen LA, Berry-Kravis EM, Berman RF, Hagerman PJ, Willemsen R, Hagerman RJ, Hukema RK. Widespread non-central nervous system organ pathology in fragile X premutation carriers with fragile X-associated tremor/ataxia syndrome and CGG knock-in mice. *Acta Neuropathol.* 2011 Oct;122(4):467-79.

[59] Todd PK, Oh SY, Krans A, He F, Sellier C, Frazer M, Renoux AJ, Chen KC, Scaglione KM, Basrur V, Elenitoba-Johnson K, Vonsattel JP, Louis ED, Sutton MA, Taylor JP, Mills RE, Charlet-Berguerand N, Paulson HL. CGG repeat-associated translation mediates neurodegeneration in fragile X tremor ataxia syndrome. *Neuron.* 2013 May 8;78(3):440-55.

[60] Ladd PD, Smith LE, Rabaia NA, Moore JM, Georges SA, Hansen RS, Hagerman RJ, Tassone F, Tapscott SJ, Filippova GN. An antisense transcript spanning the CGG repeat region of FMR1 is upregulated in premutation carriers but silenced in full mutation individuals. *Hum. Mol. Genet.* 2007 Dec 15;16(24):3174-87.
[61] Mailick MR, Hong J, Rathouz P, Baker MW, Greenberg JS, Smith L, Maenner M. Low-normal FMR1 CGG repeat length: phenotypic associations. *Front Genet.* 2014 Sep 9;5:309.
[62] Chen LS, Tassone F, Sahota P, Hagerman PJ. The (CGG)n repeat element within the 5' untranslated region of the FMR1 message provides both positive and negative cis effects on in vivo translation of a downstream reporter. *Hum. Mol. Genet.* 2003 Dec 1;12(23):3067-74. Epub 2003 Sep 30.
[63] Ramocki MB, Zoghbi HY. Failure of neuronal homeostasis results in common neuropsychiatric phenotypes. *Nature.* 2008 Oct 16;455(7215):912-8. Review.
[64] Hunter JE, Leslie M, Novak G, Hamilton D, Shubeck L, Charen K, Abramowitz A, Epstein MP, Lori A, Binder E, Cubells JF, Sherman SL. Depression and anxiety symptoms among women who carry the PM: impact of raising a child with fragile X syndrome is moderated by CRHR1 polymorphisms. *Am. J. Med. Genet. B Neuropsychiatr Genet.* 2012 Jul;159B(5):549-59.
[65] Silva F, Rodriguez-Revenga L, Madrigal I, Alvarez-Mora MI, Oliva R, Milà M. High apolipoprotein E4 allele frequency in FXTAS patients. *Genet Med.* 2013 Aug;15(8):639-42.
[66] Lozano R, Hagerman RJ, Duyzend M, Budimirovic DB, Eichler EE, Tassone F. Genomic studies in fragile X premutation carriers. *J. Neurodev Disord.* 2014;6(1):27.
[67] Hamlin A, Liu Y, Nguyen DV, Tassone F, Zhang L, Hagerman RJ. Sleep apnea in fragile X premutation carriers with and without FXTAS. Am. J. Med. Genet. B Neuropsychiatr Genet. 2011;156B (8):923–928.
[68] Leehey MA. Fragile X-associated tremor/ataxia syndrome: clinical phenotype, diagnosis, and treatment. *J. Investig. Med.* 2009 Dec;57(8):830-6. Review.
[69] Rodriguez-Revenga L, Madrigal I, Alegret M, Santos M, Milà M. Evidence of depressive symptoms in fragile-X syndrome premutated females. *Psychiatr Genet.* 2008 Aug;18(4):153-5.
[70] Bourgeois JA, Seritan AL, Casillas EM, Hessl D, Schneider A, Yang Y, Kaur I, Cogswell JB, Nguyen DV, Hagerman RJ. Lifetime prevalence of

mood and anxiety disorders in fragile X premutation carriers. *J. Clin. Psychiatry*. 2011 Feb;72(2):175-82.

[71] Hunter JE, Leslie M, Novak G, Hamilton D, Shubeck L, Charen K, Abramowitz A, Epstein MP, Lori A, Binder E, Cubells JF, Sherman SL. Depression and anxiety symptoms among women who carry the FMR1 premutation: impact of raising a child with fragile X syndrome is moderated by CRHR1 polymorphisms. *Am. J. Med. Genet. B Neuropsychiatr Genet*. 2012 Jul;159B(5):549-59.

[72] Chonchaiya W, Au J, Schneider A, Hessl D, Harris SW, Laird M, Mu Y, Tassone F, Nguyen DV, Hagerman RJ. Increased prevalence of seizures in boys who were probands with the FMR1 premutation and co-morbid autism spectrum disorder. *Hum. Genet.* 2012; 131 (4):581–589.

[73] Farzin F, Perry H, Hessl D, Loesch D, Cohen J, Bacalman S, Gane L, Tassone F, Hagerman P, Hagerman R. Autism spectrum disorders and attention-deficit/hyperactivity disorder in boys with the fragile X premutation. *J. Dev. Behav. Pediatr*. 2006; 27 (2 Suppl):S137–144.

[74] Kraan CM, Hocking DR, Georgiou-Karistianis N, Metcalfe SA, Archibald AD, Fielding J, Trollor J, Bradshaw JL, Cohen J, Cornish KM. Impaired response inhibition is associated with self-reported symptoms of depression, anxiety, and ADHD in female FMR1 premutation carriers. *Am. J. Med. Genet. B Neuropsychiatr Genet*. 2014 Jan;165(1):41-51.

[75] Bailey DB Jr, Raspa M, Olmsted M, Holiday DB. Co-occurring conditions associated with FMR1 gene variations: findings from a national parent survey. *Am. J. Med. Genet. A*. 2008; 146A (16): 2060–2069.

[76] Hagerman RJ, Coffey SM, Maselli R, Soontarapornchai K, Brunberg JA, Leehey MA, Zhang L, Gane LW, Fenton-Farrell G, Tassone F, Hagerman PJ. Neuropathy as a presenting feature in fragile X-associated tremor/ataxia syndrome. *Am. J. Med. Genet. A*. 2007 Oct 1;143A(19):2256-60.

In: Allelic Forms of the FMR1 Gene
Editor: Montserrat Milà
ISBN: 978-1-63321-914-4

Chapter 6

Genetic Counseling of *FMR1*

I. Madrigal
Biochemistry and Molecular Genetics Department,
Hospital Clínic and IDIBAPS
Centre for Biomedical Research on Rare Diseases (CIBERER),
ISCIII, Barcelona, Spain
IDIBAPS (Institut d'Investigacions Biomèdiques August Pi i Sunyer).
Barcelona, Spain

Abstract

Fragile X syndrome is the most common form of inherited intellectual disability, with an estimated incidence of 1 in 4,000 males and 1 in 6,000 females. Each diagnosis of an *FMR1* mutation has far reaching clinical and reproductive implications for the extended family. Until now genetic counseling was offered based on the expansion risk in premutation carrier women, but the description of *FMR1*-associated disorders has increased the complexity of the genetic counseling for FXS families, especially *FMR1* premutation carriers. Male individuals carrying full mutated alleles present with intellectual disability while the penetrance is incomplete in females (30-50%). Premutation allele carriers are intellectually unaffected, but several *FMR1* premutation-related disorders have been described. The most prevalent are fragile X-associated primary ovarian insufficiency and fragile X-associated tremor/ataxia syndrome, but behavioral features such as impaired executive function, social deficits or anxiety have also been related to several *FMR1* premutation carriers. Premutation women have 50% of risk of transmitting a premutation/full mutation allele to their offspring, depending on the CGG expansion repeat and the presence of AGG interruptions. On the contrary, premutation men carriers will only transmit the premutation allele to their daughters. Some issues such as risk assessment for intermediate alleles and the clinical prognosis for females with full mutations still remain challenging. Genetic counselors must have an updated and solid understanding of this genetic condition and the *FMR1*-associated disorders in order to cover all the counseling aspects of these disorders.

Keywords: *FMR1,* FXS, premutation alleles, full mutation alleles, genetic counseling, intermediate alles

Introduction

The World Health Organization defines Genetic Counseling as the process through which trained genetic counselors inform about the genetic aspects of a particular genetic disease to those who are at an increased risk of developing the disorder or of passing it on to their unborn offspring. A genetic counselor must provide accurate information to the patient and their families about the clinical consequences of the conditions, the inheritance of illnesses and their risks of recurrence, management and/or treatment, quality of life, life expectancy and social aspects. These professionals must then work in a wide variety of fields such as general genetics, prenatal care or family planning. The counseling process consists of two stages: pre and the post-test genetic counseling. In pre-test genetic counseling individuals are informed about the purpose of the test, the clinical consequences of the conditions, including the phenotypic features, inheritance patterns, the reliability and limitations of the test and the possible psychological impacts to the counselees and their relatives.

In post-test genetic counseling, after disclosure of the test results, the counselor should focus on the emotional impact on the counselee and the relatives involved. The counselees must be informed about implications to them and their close relatives, including management and/or treatment options, quality of life, life expectancy and available reproductive choices. Counselors must ensure the privacy and confidentiality of the results.

Genetic Counseling in *FMR1*-Associated Disorders

Fragile X syndrome (FXS, #300624) is the most common form of inherited intellectual disability (ID), with an estimated incidence of 1 in 4,000 males and 1 in 6,000 females [1]. The molecular basis of this syndrome is mainly the expansion of an unstable CGG repeat in the 5' untranslated region of the fragile X syndrome gene (*FMR1*). The polymorphic CGG repeat of the *FMR1* gene is distributed in the population in four allelic classes according to repeat length (Table 1). Alleles ranging from 6 to 44 CGG repeats are the most common in the general population and have a stable transmission to the next generation. Alleles within the 45–54 CGG repeat range are called intermediate alleles and most are stable. Individuals carrying intermediate alleles are phenotypically normal. Alleles about 55–200 repeats are called premutation alleles. These alleles are associated with a significant elevation of *FMR1* mRNA levels [2-3] and are highly unstable during maternal transmission. Premutation carrier individuals do not present with ID, but they have a risk of developing fragile X-associated tremor/ataxia syndrome (FXTAS), premature ovarian insufficiency (FXPOI) or other psychological symptoms [4-7]. Finally, expansion of the repeat region to more than 200 CGG trinucleotide sequences is called full mutation. *FMR1* is an X-linked gene regulated by methylation of the promoter during X inactivation in somatic cells of females. This leads to gene silencing and insufficient synthesis of the *FMR1* protein (FMRP). Expansions over 200 CGGs lead to hypermethylation of the CpG island resulting in non-expression of the *FMR1* gene and absence of the FMRP. The lack of FMRP is the direct cause of the FXS phenotype [8]. Besides CGG repeat expansions, it has been reported that almost 1% of individuals with FXS have a partial or full deletion, or point mutation of the *FMR1* gene [9-12].

Until now genetic counseling was offered based on the expansion risk in premutation carrier women, but the description of *FMR1*-associated disorders has increased the complexity of the genetic counseling for FXS families, particularly to *FMR1* premutation carriers. Furthermore, some issues such as risk assessment for intermediate alleles and the clinical prognosis for females with full mutations still remain challenging.

Most of the families who first receive a diagnosis of FXS have no prior knowledge of *FMR1*-premutation disorders, and it is essential that individuals under study receive pre and post genetic counseling, including information about possible outcomes and the implications of carrying full, premutation or intermediate alleles.

Genetic counseling of FXS requires attention to a wide range of clinical manifestations including developmental, neurodegenerative, and reproductive symptoms that may vary in age of onset and severity. Counselors must have a solid understanding of this genetic condition, including the trinucleotide repeat instability and the phenotypic variability. Table 1 shows associated *FMR1* allele phenotypes.

Table 1. *FMR1* alleles and associated phenotypes

Allele	CGG repeat range	Phenotype
Normal	5-44	Normal
Intermediate	45-54	Normal
Premutation	55–200	Normal, FXTAS, FXPOI and others
Full Mutation	>200 (methylated)	ID in 100% males and reduced penetrance in females (30-50%)

Assessment Risk Based on CGGs Expansion

Normal Range and Intermediate Alleles

Alleles ranging from 6 to 44 CGG repeats, the most common in the general population, and most of the intermediate alleles have stable transmission to the next generation. Intermediate alleles may show some instability, including expansion to a full mutation in two generations [13-15].

Premutation Alleles

Premutation alleles are highly unstable during maternal transmission and may expand to a higher CGG repeat size or even to a full mutation in only one generation. Even expansion of paternal premutation alleles is possible; the expansion from a premutation allele to a full mutation has only been observed through female meioses.

Thus, all the daughters from a premutation male will inherit the premutation allele and will not manifest FXS. Table 2 shows the risk of expansion of the different allele types. The expansion risk of a *FMR1* allele depends both on CGG repeat size and the presence of AGG interruptions. In 1994, Eichler et al. suggested that AGGs interspersed within the *FMR1* repeat region may be linked to repeat stability and the risk of expansion [16]. In the general population, almost 95% of alleles have one or two AGG interruptions, which the most common allele pattern is two AGGs at positions 10 or 11 and 20 or 21 repeats. In contrast, alleles in FXS families contain no or few AGGs at the 5′ end and they contain long stretches of uninterrupted CGGs at the 3′ end [16,17]. Presumably maternal alleles with no AGG interruptions confer increased risk for unstable transmissions and thus, the inclusion of AGG genotype studies would be of benefit in clinical practice. AGG testing is expected to reassure premutation carriers with AGG interruptions while alerting premutation carriers with alleles without AGG interruptions who would have the highest risk for instability.

Table 2. Risk of expansion of FMR1 alleles

Allele	CGG repeat range	Risk of expansion to a full mutation for females*
Normal	5-44	0%
Intermediate	45-54	0%
Premutation	55–59 60–69 70–79 80–89 90–99 >100	3.70 % 5.30 % 31.10 % 57.80 % 80.10 % 94–100 %
Full Mutation	>200 (methylated)	

*[68-70].

Full Mutation Alleles

Full mutated males never transmit the full mutated allele to their daughters, only an allele in the premutation range. Full mutated females have a 50% of transmission risk of the full mutated allele to their offspring: all male descendants carrying the full mutation allele will be affected while a smaller percentage of women will present the syndrome.

Genetic Counseling Based on *FMR1* Alleles

Intermediate Alleles

Intermediate alleles present a 45–54 CGG repeat range [18] and have a frequency in the general population of 1/35 to 1/57 females [19,20]. Intermediate allele carriers do not manifest ID, FXTAS or FXPOI, although some groups have suggested associations between intermediate alleles and some disorders such as Parkinson's disease [21], primary ovarian insufficiency [22], and autism and cognitive disabilities [23,24]. However, these findings have not consistently been supported by other studies [25-27]. A frequent concern in FXS screening is the genetic counseling to intermediate allele carriers. Currently, the follow-up of individuals with no AGG interruptions is indicated, despite the low risk of expansion in the next generation.

Premutation Alleles

Individuals carrying premutation alleles are intellectually unaffected, but in the last decades several premutation-related disorders have been described. Among them the most prevalent are FXPOI and FXTAS [28,29]. In addition, behavioral features such as impaired executive function, social deficits or anxiety have been related to some *FMR1* premutation carriers [30]. Premutation alleles contain trinucleotide expansions in the range of 55 to 200 CGG repeats. In Western populations premutation frequencies in females range from 1/151 to 1/259, but there is some evidence of ethnic and racial variability. In Israel, for example, the premutation frequency is around 1/113 and in Taiwan 1/837 [20,31-33]. Some authors state that the different reproductive aptitudes of female premutation carriers and the differences in mean maternal age may contribute to this variation in premutation frequencies [34]. In males the rate ranges from 1/468 to 1/813 [20,35].

Men carrying the premutation allele must mainly be counseled about the risk of FXTAS. This syndrome presents with incomplete penetrance even among premutation carriers with identical CGG-repeat lengths. Approximately 50% of premutation carrier males will develop neurodegenerative symptoms of FXTAS after the age of 50 [36]. The first clinical signs of the syndrome typically appear when patients are in their 50s and 60s, with a mean tremor onset at approximately 60 years and ataxia onset at 62 years [37]. The risk and severity of the disorder appear to be related to the CGG repeat size, with higher risk in larger repeats [38]; nevertheless, a biomarker that predicts the apparition of FXTAS or the protective factors in asymptomatic carriers has still not been identified. Other clinical signs associated with premutation male carriers are neuroendocrine dysfunction, including testosterone deficiency [39], hypertension [40] or bowel and urinary incontinence [41].

Genetic counseling of premutation carrier women must be addressed to the risk for premature ovarian insufficiency, besides the risk for expansion and the FXTAS syndrome. In women, FXTAS is much less common than in men, as many as 16% will develop FXTAS symptoms, it presents a later age of onset and is milder in presentation. The risk for FXPOI in these women increases around 20% with an age of onset before the age of 40 years [28]. Other pathologies associated with females premutation carriers are psychiatric disorders such as depression, anxiety or mood disorders [42-44], migraine [45], immune-mediated disorders, particularly hypothyroidism (15.9%) [36], and fibromyalgia (25%) [7,46,47]. The latter two in particular are even more common among women with FXTAS, with a frequency of 43% and 50% , respectively [7].

Full Mutation Alleles

Full mutations are responsible for FXS, a spectrum of clinical features which includes physical, cognitive and behavioral aspects. All males carrying the full mutation will present with mild to severe intellectual disability and exhibit a variety of maladaptive behaviors overlapping those described for autism spectrum disorders [48,49]. On the contrary, only 50%-70% of women manifest FXS symptoms, albeit in a milder form than men, and some appear to be completely unaffected or exhibit minor neurobehavioral features [50-52]. Despite a common genetic etiology, the clinical presentation of FXS is variable and this variability is related to residual levels of the FMRP due to the presence of CGG expansion size mosaicism, different methylation levels of the full mutation allele and X-inactivation that leads to a differential pattern of FMRP expression within tissues [51].

On the other hand, there are full mutation carrier males with a standardized IQ score of 70 or higher who are known as high functioning males. In normal males, the 5' CpG island containing the CGG trinucleotide repeat is not methylated, whereas a CGG triplet repeat expans to more than ~200 repeats, this site is methylated and the *FMR1* is transcriptionally silenced

[53]. Many mildly affected individuals show mosaic methylation at the *FMR1* promoter.

Regarding their offspring, their daughters will inherit an allele in the premutation range. High functioning males do not present with intellectual disability but recently, FXTAS has been described in a high functioning male [54] and in the case of an unmethylated mosaic (premutation-full mutation) male [55]. The latter suggests that the definition of FXTAS should also include those cases with an expanded allele the size and lack of methylation of which leads to RNA toxicity.

FMR1 Point Mutations/Deletions

It has been suggested that patients with a clinical FXS-like phenotype but not carrying the *FMR1* gene full mutation should be screened for *FMR1* mutations [56,57]. It is estimated that *FMR1* point mutations or deletions are responsible for up 1% of FXS cases. Nevertheless, the prevalence of mutations in the *FMR1* coding region is still not well known since the standard FXS protocols only comprise the study of the CGG repeats size. These mutations can be *de novo* or inherited from a carrier mother. A male carrying a point mutation or deletion will always have carrier daughters, who might be affected depending on the X-chromosome inactivation. Carrier females have 50% of risk of transmitting the mutated allele. Within this 50%, all males will be affected and females will be variably affected, depending on the X-inactivation.

Reproductive Options for Premutation/Full Mutation Carriers

Individuals at risk for passing on FXS mutations to their offspring have a variety of pre- and postconception options available. Regardless of the option they choose, women must be advised of any contraindication of these procedures [58]. When offering the reproductive options to a couple at risk, a fertility issue has to be considered since the risk of POF has significant reproductive implications. First of all, the onset of POF is difficult to predict. Secondly, the subtle endocrine perturbations (elevated follicle-stimulating hormone levels) observed in these women decreases the efficiency of ovarian stimulation required for preimplantational diagnosis and finally, there is evidence that premutation carriers may have hormonal changes suggestive of early ovarian aging despite regular menstrual cycles (See Chapter 5).

Prenatal Diagnosis for FXS

Prenatal diagnosis for FXS is possible using chorionic villi or amniotic fluid cells and it depends on several factors, including the physician performing the procedure and patient preference. When the establishment of the sizing of the expansion is sufficient to determine the genotype, the prenatal diagnosis on chorionic villi allows performing a therapeutic abortion at early stages in the pregnancy. On the contrary, the methylation pattern of a full mutation is established after the 14th week of pregnancy, thus the use of a methylation-sensitive method (i.e Southern Blot) is not suitable for early prenatal diagnosis on DNA from chorionic villi [59]. Moreover, the risk of miscarriage may be lower on amniotic fluid cell sampling.

Table 3. Genetic counseling in *FMR1*

Progenitor allele and gender	% expected offspring	Offspring outcome	
		Males	Females
Normal male	No risk	Normal	Normal
Normal female	No risk	Normal	Normal
IA carrier male	Females IA range	Normal	Normal
IA carrier female	50% NA range 50% IA range	Normal	Normal
PM carrier male	Females PM range	Normal	Normal, FXTAS, FXPOI and others
PM carrier female	50% NA range 50%* PM/ FM*	Normal Normal, FXTAS, and others 100% ID	Normal Normal, FXTAS, FXPOI and others 30-50% ID
FM carrier male	females PM range	Normal	Normal, FXTAS, FXPOI and others
FM carrier female	50% NA range 50% FM	Normal 100% ID	Normal 30-50% ID

NA: normal allele, IA: intermediate allele; PM: premutation allele; FM: full mutation allele.

*percentage varies according to the expansion risk indicated in Table 2.

Prenatal diagnosis should be offered to women with premutation or full mutations, but it is not intended for the pregnant partner of a premutation male carrier. These males, nevertheless, should receive genetic counseling about phenotypic risk to their daughters, who will inherit the premutation allele. Neither is prenatal testing intended for intermediate allele carriers. As indicated before, there are no reports of intermediate alleles expanding to full mutations in a single generation. The risk for a female premutation carrier of transmitting the expansion is 50%, and the risk for the maternal premutation to expand to full mutation is proportional to its size and AGG content. Table 3 summarizes the possible outcomes depending on the carrier state of each parent.

The identification of a female fetus with the fragile X full mutation entails a significant challenge for professionals. The probability that a full mutation carrier fetus is affected is around 50-70%, but at present there are no biomarkers to predict this affectation. The diagnosis of a FXS patient requires extending the study to the mother in order to confirm her premutation carrier status.

Preimplantational Diagnosis for FXS

Preimplantation genetic diagnosis (PGD) is a technique based on the genetic analysis of an embryo obtained through in vitro fecundation. The most common option is to perform the genetic study on a day-3 embryo which is then transferred to the uterus. This approach increases the risk of embryo viability due to the biopsy. Further, the possibility of embryo mosaicism has been described at the cleavage stage and self-correction of aneuploidies between the cleavage and blastocyst stages [60]. Polar body biopsies can be used to avoid any misdiagnosis due to embryo mosaicism, although only maternal genetic information is obtained. The third option is to biopsy trophectoderm cells from blastocysts (5-6 day embryo), which is less invasive and has higher concordance between inner cell mass and trophectoderm cells [61]. PGD has several advantages over prenatal diagnosis. The diagnosis is performed in the embryo, avoiding the parental stress and the emotional trauma of a termination of pregnancy.

Offspring Renouncement and Adoption

These two options are drastic and are usually rejected by the couples. Therefore they are currently seldom observed as the first choice option by couples at risk that wish to prevent the birth of an affected child.

Donor Germline Cell

Another possible option is egg or sperm donation for female and male mutation carriers, respectively. Based on our experience, around 10% of FXS families chose for a germline cell donation [62]. At present, it is a very good option for premutation carrier females who have been involved in previous prenatal diagnosis and termination of pregnancies. Moreover, given the relatively high prevalence of FXS in the general population, potential gamete donors should be tested for this syndrome.

FMR1 Testing Guidelines

Several scientific societies such as the European Molecular genetics Quality Network (EMQN), the American College of Medical Genetics (ACMG), the National Society of genetic Counselors (NSGC) and the American College of Obstetrics and Gynecology (ACOG) have published best practice guidelines for the molecular genetic testing and diagnosis of FXS and other fragile X-associated disorders [18,63-66]. The best practice guidelines for genetic analysis and reporting in FXS, FXPOI, and FXTAS are listed in Table 4. Appropriate *FMR1* molecular testing is very important for optimal genetic counseling in the fragile X-associated disorders due to the particular pattern and transmission of the CGG repeat. Although not being endorsed by current guidelines, screening women without known risk factors to FXS is increasingly being offered. In 2005 Musci and Caughey demonstrated the efficacy and cost effectiveness of prenatal population-based fragile X carrier screening [67].

Table 4. *FMR1* mutation testing recommendations

FMR1 mutation testing is recommended for*:
Individuals of either sex with intellectual disability, developmental delay or autism
Individuals who have a family history of FXS
Women with a family history of *FMR1*-related disorders, including FXPOI
Women with reproductive or fertility problems associated with elevated levels of follicle stimulating hormone (FSH)
Individuals with late onset tremor or cerebellar ataxia of unknown origin
Given the relatively high prevalence of FXS in the general population, potential gamete donors should be tested for this syndrome

*Sherman, 2005; ACOG, 20010.

Conclusion

When first described, FXS was a rare X-linked disease responsible for intellectual disabilities in males and females in a lesser percentage. Research into *FMR1* has brought to light the molecular and phenotypic complexity of *FMR1*-related diseases. The first issue to consider is that a FXS patient will always have a premutation carrier mother who should receive genetic counseling.

Premutation carrier females have a 50% risk of transmitting a premutation/full mutated allele to their offspring, unlike carrier males who will only transmit the premutation allele. The expansion risk in females depends on the CGG repeats number and the AGG interruptions. Lastly, premutation carriers are at risk not only of having FXS offspring (females) but also of FXPOI (females), FXTAS and other disorders. The complexity of the issues surrounding genetic testing and the management of *FMR1*-associated disorders has increased, and investigation related to this gene covers many aspects, including molecular factors, epigenetic factors, emotional issues or targeted pharmaceuticals.

As the knowledge regarding disease causing mechanisms evolves, genetic counselors must have an updated and solid understanding of this genetic condition and the *FMR1*-associated disorders in order to cover all the counseling aspects of these syndromes.

References

[1] Hagerman RJ. The physical and behavioural phenotype, in: RJ Hagerman, P Hagerman (Eds.), Fragile-X Syndrome: Diagnosis, Treatment and Research, The Johns Hopkins University Press, Baltimore, 2002, pp. 3–109.

[2] Tassone F, Hagerman RJ, Loesch DZ, Lachiewicz A, Taylor AK, Hagerman PJ. Fragile X males with unmethylated, full mutation trinucleo tide repeat expansions have elevated levels of FMR1 messenger RNA. *Am. J. Med. Genet*. 2000a;94:232–236.

[3] Tassone F, Hagerman RJ, Taylor AK, Gane LW, Godfrey TE, Hagerman PJ. Elevated levels of FMR1 mRNA in carrier males: A new mechanism of involvement in the fragile-X syndrome. *Am. J. Hum. Genet.* 2000b;66:6-15.

[4] Hagerman RJ, Leehey M, Heinrichs W, Tassone F, Wilson R, Hills J, Grigsby J, Gage B, Hagerman PJ. Intention tremor, parkinsonism, and generalized brain atrophy in male carriers of fragile X. *Neurology* 2001;57:127-130.

[5] Jacquemont S, Hagerman RJ, Leehey M, Grigsby J, Zhang L, Brunberg JA, Greco C, Des Portes V, Jardini T, Levine R, Berry-Kravis E, Brown WT, Schaeffer S, Kissel J, Tassone F, Hagerman PJ. Fragile X premutation tremor/ataxia syndrome: molecular, clinical, and neuroimaging correlates. *Am. J. Hum. Genet.* 2003;72:869-878.

[6] Hessl D, Tassone F, Loesch DZ, Berry-Kravis E, Leehey MA, Gane LW, Barbato I, Rice C, Gould E, Hall DA, Grigsby J, Wegelin JA, Harris S, Lewin F, Weinberg D, Hagerman PJ, Hagerman RJ. Abnormal elevation of FMR1 mRNA is associated with psychological symptoms in individuals with the fragile X premutation. *Am. J. Med. Genet. B Neuropsychiatr. Genet.* 2005;139B:115-21.

[7] Coffey SM, Cook K, Tartaglia N, Tassone F, Nguyen DV, Pan R, Bronsky HE, Yuhas J, Borodyanskaya M, Grigsby J, Doerflinger M, Hagerman PJ, Hagerman RJ. Expanded clinical phenotype of women with the FMR1 premutation. *Am. J. Med. Genet. A.* 2008;146A:1009-16.

[8] de Vries BB, Mohkamsing S, van den Ouweland AM, Mol E, Gelsema K, van Rijn M, Tibben A, Halley DJ, Duivenvoorden HJ, Oostra BA, Niermeijer MF. Screening with the FMR1 protein test among mentally retarded males. *Hum. Genet.* 1998;103:520-522.

[9] De Boulle K, Verkerk AJ, Reyniers E, Vits L, Hendrickx J, Van Roy B, Van den Bos F, de Graaff E, Oostra BA, Willems PJ. A point mutation

in the FMR-1 gene associated with fragile X mental retardation. *Nat. Genet.* 1993;3:31–5.

[10] Lugenbeel KA, Peier AM, Carson NL, Chudley AE, Nelson DL. Intragenic loss of function mutations demonstrate the primary role of FMR1 in fragile X syndrome. *Nat. Genet. 1995*;10:483–5.

[11] Strom CM, Crossley B, Redman JB, Buller A, Quan F, Peng M, McGinnis M, Fenwick RG Jr, Sun W. Molecular testing for fragile X syndrome: lessons learned from 119,232 tests performed in a clinical laboratory. *Genet Med. Off J. Am. Coll. Med. Genet* 2007;9:46e51.

[12] Handt M, Epplen A, Hoffjan S, Mese K, Epplen JT, Dekomien G. Point mutation frequency in the FMR1 gene as revealed by fragile X syndrome screening. *Mol. Cell Probes*. 2014 Aug 27. pii: S0890-8508(14)00042-5. doi: 10.1016/j.mcp.2014.08.003.

[13] Terracciano A, Pomponi MG, Marino GM, Chiurazzi P, Rinaldi MM, Dobosz M, Neri G. Expansion to full mutation of a FMR1 intermediate allele over two generations. *Eur. J. Hum. Genet.* 2004;12:333-336.

[14] Zuniga A, Juan J, Mila M, Guerrero A. Expansion of an intermediate allele of the *FMR1* gene in only two generations. *Clin. Genet* 2005;68:471–473.

[15] Fernandez-Carvajal I, Lopez Posadas B, Pan R, Raske C, Hagerman PJ, Tassone F. Expansion of an FMR1 grey-zone allele to a full mutation in two generations. *J. Mol. Diagn* 2009;11:306-310.

[16] Eichler EE, Holden JJ, Popovich BW, Reiss AL, Snow K, Thibodeau SN, Richards CS, Ward PA, Nelson DL. Length of uninterrupted CGG repeats determines instability in the FMR1 gene. *Nat. Genet.* 1994;8:88-94.

[17] Eichler EE, Macpherson JN, Murray A, Jacobs PA, Chakravarti A, Nelson DL. Haplotype and interspersion analysis of the FMR1 CGG repeat identifies two different mutational pathways for the origin of the fragile X syndrome. *Hum. Mol. Genet*. 1996;5:319-30.

[18] Sherman S, Pletcher BA, Driscoll DA. Fragile X syndrome: diagnostic and carrier testing. *Genet Med*. 2005;7:584-7.

[19] Cronister A, Teicher J, Rohlfs EM, Donnenfeld A, Hallam S. Prevalence and instability of fragile X alleles: implications for offering fragile X prenatal diagnosis. *Obstet. Gynecol*. 2008;111:596-601.

[20] Seltzer MM, Baker MW, Hong J, Maenner M, Greenberg J, Mandel D. Prevalence of CGG expansions of the FMR1 gene in a US population-based sample. *American Journal of Medical Genetics Part B* 2012;159B:589–597.

[21] Loesch DZ, Khaniani MS, Slater HR, Rubio JP, Bui QM, Kotschet K, D'Souza W, Venn A, Kalitsis P, Choo AK, Burgess T, Johnson L, Evans A, Horne M. Small CGG repeat expansion alleles of FMR1 gene are associated with parkinsonism. *Clin. Genet.* 2009a;76:471-6.

[22] Bodega B, Bione S, Dalprà L, Toniolo D, Ornaghi F, Vegetti W, Ginelli E, Marozzi A. Influence of intermediate and uninterrupted FMR1 CGG expansions in premature ovarian failure manifestation. *Hum. Reprod.* 2006;21:952-7.

[23] Aziz M, Stathopulu E, Callias M, Taylor C, Turk J, Oostra B, Willemsen R, Patton M. Clinical features of boys with fragile X premutations and intermediate alleles. *Am. J. Med. Genet. B Neuropsychiatr. Genet.* 2003;121B:119-27.

[24] Loesch DZ, Godler DE, Khaniani M, Gold E, Gehling F, Dissanayake C, Burgess T, Tassone F, Huggins R, Slater H, Choo KH. Linking the FMR1 alleles with small CGG expansions with neurodevelopmental disorders: preliminary data suggest an involvement of epigenetic mechanisms. *Am. J. Med. Genet. A.* 2009b;149A:2306-10.

[25] Ennis S, Murray A, Youings S, Brightwell G, Herrick D, Ring S, Pembrey M, Morton NE, Jacobs PA. An investigation of FRAXA intermediate allele phenotype in a longitudinal sample. *Ann. Hum. Genet.* 2006;70:170-80.

[26] Bennett CE, Conway GS, Macpherson JN, Jacobs PA, Murray A. Intermediate sized CGG repeats are not a common cause of idiopathic premature ovarian failure. *Hum. Reprod.* 2010;25:1335-8.

[27] Madrigal I, Xunclà M, Tejada MI, Martínez F, Fernández-Carvajal I, Pérez-Jurado LA, Rodriguez-Revenga L, Milà M. Intermediate FMR1 alleles and cognitive and/or behavioural phenotypes. *Eur. J. Hum. Genet.* 2011;19:921-3.

[28] Sherman, S. L. Premature ovarian failure in the fragile X syndrome. *Am. J. Med. Genet.* 2000;97:189–194.

[29] Garcia-Arocena, D., Hagerman, P. J.. Advances in understanding the molecular basis of FXTAS. *Human Molecular Genetic* 2010;19:R83–89.

[30] Bailey DB Jr, Raspa M, Olmsted M, Holiday DB. Co-occurring conditions associated with FMR1 gene variations: findings from a national parent survey. *Am. J. Med. Genet A.* 2008;146A:2060-9.

[31] Rousseau, F., Rouillard, P., Khandjian, E. W., & Morgan, K. Prevalence of carriers of premutation-size alleles of the FMR1 gene and implications for the population genetics of the fragile X syndrome. *American Journal of Human Genetics* 1995;57:1006–1018.

[32] Toledano-Alhadef H, Basel-Vanagaite L, Magal N, Davidov B, Ehrlich S, Drasinover V, Taub E, Halpern GJ, Ginott N, Shohat M. Fragile-X carrier screening and the prevalence of premutation and full-mutation carriers in Israel. *American Journal of Human Genetics* 2001;69:352–360.

[33] Tzeng, C. C., Tsai, L. P., Hwu, W. L., Lin, S. J., Chao, M. C., Jong, Y. J., Chu, S. Y., Chao, W. C., and Lu, C. L. Prevalence of the FMR1 mutation in Taiwan assessed by large-scale screening of newborn boys and analysis of DXS548-FRAXAC1 haplotype. *Am. J. Med. Genet.* 2005;133A:37–43.

[34] Genereux DP, Laird CD. Why do fragile X carrier frequencies differ between Asian and non-Asian populations? *Genes Genet. Syst.* 2013;88:211-24.

[35] Dombrowski C, Lévesque S, Morel ML, Rouillard P, Morgan K, Rousseau F. Premutation and intermediate-size FMR1 alleles in 10572 males from the general population: loss of an AGG interruption is a late event in the generation of fragile X syndrome alleles. *Human Molecular Genetics* 2002;11:371–378.

[36] Rodriguez-Revenga L, Madrigal I, Pagonabarraga J, Xunclà M, Badenas C, Kulisevsky J, Gomez B, Milà M. Penetrance of FMR1 premutation associated pathologies in fragile X syndrome families. *Eur. J. Hum. Genet.* 2009;17:1359-62.

[37] Leehey MA, Berry-Kravis E, Min SJ, Hall DA, Rice CD, Zhang L, Grigsby J, Greco CM, Reynolds A, Lara R, Cogswell J, Jacquemont S, Hessl DR, Tassone F, Hagerman R, Hagerman PJ. Progression of tremor and ataxia in male carriers of the FMR1 premutation. *Mov. Disord.* 2007;22:203-6.

[38] Leehey MA, Berry-Kravis E, Goetz CG, Zhang L, Hall DA, Li L, Rice CD, Lara R, Cogswell J, Reynolds A, Gane L, Jacquemont S, Tassone F, Grigsby J, Hagerman RJ, Hagerman PJ. FMR1 CGG repeat length predicts motor dysfunction in premutation carriers. *Neurology.* 2008;70:1397-402.

[39] Greco CM, Soontrapornchai K, Wirojanan J, Gould JE, Hagerman PJ, Hagerman RJ. Testicular and pituitary inclusion formation in fragile X associated tremor/ataxia syndrome. *J. Urol.* 2007;177:1434-7.

[40] Hamlin AA, Sukharev D, Campos L, Mu Y, Tassone F, Hessl D, Nguyen DV, Loesch D, Hagerman RJ. Hypertension in FMR1 premutation males with and without fragile X-associated tremor/ataxia syndrome (FXTAS). *Am. J. Med. Genet. A.* 2012;158A:1304-9.

[41] Hagerman PJ, Hagerman RJ. Fragile X-associated tremor/ataxia syndrome--an older face of the fragile X gene. *Nat. Clin. Pract. Neurol.* 2007;3:107-12.

[42] Rodriguez-Revenga L, Madrigal I, Alegret M, Santos M, Milà M. Evidence of depressive symptoms in fragile-X syndrome premutated females. *Psychiatr Genet.* 2008;18:153-5.

[43] Bourgeois JA, Seritan AL, Casillas EM, Hessl D, Schneider A, Yang Y, Kaur I, Cogswell JB, Nguyen DV, Hagerman RJ. Lifetime prevalence of mood and anxiety disorders in fragile X premutation carriers. *J. Clin. Psychiatry.* 2011;72:175-82.

[44] Hunter JE, Sherman S, Grigsby J, Kogan C, Cornish K. Capturing the fragile X premutation phenotypes: a collaborative effort across multiple cohorts. *Neuropsychology.* 2012;26:156-64.

[45] Au J, Akins RS, Berkowitz-Sutherland L, Tang HT, Chen Y, Boyd A, Tassone F, Nguyen DV, Hagerman R. Prevalence and risk of migraine headaches in adult fragile X premutation carriers. *Clin. Genet.* 2013;84:546-51.

[46] Leehey MA, Legg W, Tassone F, Hagerman R. Fibromyalgia in fragile X mental retardation 1 gene premutation carriers. *Rheumatology* (Oxford). 2011;50:2233-6.

[47] Rodriguez-Revenga L, Madrigal I, Blanch-Rubió J, Elurbe DM, Docampo E, Collado A, Vidal J, Carbonell J, Estivill X, Mila M. Screening for the presence of FMR1 premutation alleles in women with fibromyalgia. *Gene.* 2013;512:305-8.

[48] Farzin F, Perry H, Hessl D, Loesch D, Cohen J, Bacalman S, Gane L, Tassone F, Hagerman P, Hagerman R. Autism spectrum disorders and attention-deficit/hyperactivity disorder in boys with the fragile X premutation. *J. Dev. Behav. Pediatr.* 2006;27:S137-44.

[49] Boyle L, Kaufmann WE. The behavioral phenotype of FMR1 mutations. *Am. J. Med. Genet. C. Semin. Med. Genet.* 2010 Nov 15;154C:469-76.

[50] Cronister A, Schreiner R, Wittenberger M, Amiri K, Harris K, Hagerman RJ. Heterozygous fragile X female: historical, physical, cognitive, and cytogenetic features. *Am. J. Med. Genet.* 1991;38:269-74.

[51] de Vries BB, Jansen CC, Duits AA, Verheij C, Willemsen R, van Hemel JO, van den Ouweland AM, Niermeijer MF, Oostra BA, Halley DJ. Variable FMR1 gene methylation of large expansions leads to variable phenotype in three males from one fragile X family. *J. Med. Genet.* 1996;33:1007-10.

[52] Keysor CS, Mazzocco MM. A developmental approach to understanding Fragile X syndrome in females. *Microsc. Res. Tech*. 2002;57:179-86.

[53] Hornstra I.K., Nelson D.L., Warren S.T., Yang T.P. High resolution methylation analysis of the FMR-1 gene trinucleotide repeat region in Fragile X syndrome. *Hum. Mol. Genet*. 1993;2:1659-1665

[54] Loesch DZ, Sherwell S, Kinsella G, Tassone F, Taylor A, Amor D, Sung S, Evans A. Fragile X-associated tremor/ataxia phenotype in a male carrier of unmethylated full mutation in the FMR1 gene. *Clin. Genet.* 2012;82:88-92

[55] Santa María L, Pugin A, Alliende MA, Aliaga S, Curotto B, Aravena T, Tang HT, Mendoza-Morales G, Hagerman R, Tassone F. FXTAS in an unmethylated mosaic male with fragile X syndrome from Chile. *Clin. Genet.* 2014;86:378-82

[56] Grønskov K, Brøndum-Nielsen K, Dedic A, Hjalgrim H. A nonsense mutation in FMR1 causing fragile X syndrome. *Eur. J. Hum. Genet.* 2011;19:489e91.

[57] Myrick LK, Nakamoto-Kinoshita M, Lindor NM, Kirmani S, Cheng X, Warren ST. Fragile X syndrome due to a missense mutation. *Eur. J. Hum. Genet.* 2014;22:1185-9.

[58] Mor YS, Schenker JG. Ovarian Hyperstimulation Syndrome and Thrombotic Events. *Am. J. Reprod. Immunol.* 2014 Aug 22. doi: 10.1111/aji.12310.

[59] Devys D, Biancalana V, Rousseau F, Boue J, Mandel JL, Oberle I. Analysis of full fragile X mutations in fetal tissues and monozygotic twins indicate that abnormal methylation and somatic heterogeneity are established early in development. *Am. J. Med. Genet* 1992;43:208–216.

[60] Frumkin T,MalcovM, Yaron Y, Ben-Yosef D. Elucidating the origin of chromosomal aberrations in IVF embryos by preimplantation genetic analysis. *Mol. Cell Endocrinol*. 2008;282:112–9.

[61] Capalbo A, Wright G, Themaat L, Elliott T, Rienzi L, Nagy ZP. FISH reanalysis of inner cell mass and trophectoderm samples of previously array-CGH screened blastocysts reveals high accuracy of diagnosis and no sign of mosaicism or preferential allocation. *Hum. Reprod.* 2013;28:2298-307.

[62] Xunclà M, Badenas C, Domínguez M, Rodríguez-Revenga L, Madrigal I, Jiménez L, Soler A, Borrell A, Sánchez A, Milà M. Fragile X syndrome prenatal diagnosis: parental attitudes and reproductive responses. *Reprod. Biomed. Online*. 2010;21:560-5.

[63] Kronquist KE, Sherman SL, Spector EB. Clinical significance of tri-nucleotide repeats in Fragile X testing: a clarification of American College of Medical Genetics guidelines. *Genet Med.* 2008;10:845-7.

[64] American College of Obstetricians and Gynecologists Committee on Genetics. ACOG Committee Opinion No. 469: Carrier screening for fragile X syndrome. *Obstet. Gynecol.* 2010;116:1008-10.

[65] Finucane B, Abrams L, Cronister A, Archibald AD, Bennett RL, McConkie-Rosell A. Genetic counseling and testing for FMR1 gene mutations: practice guidelines of the national society of genetic counselors. *J. Genet Couns.* 2012;21:752-60.

[66] Biancalana V, Glaeser D, McQuaid S, Steinbach P. EMQN best practice guidelines for the molecular genetic testing and reporting of fragile X syndrome and other fragile X-associated disorders. *Eur. J. Hum. Genet.* 2014 Sep 17. doi: 10.1038/ejhg.2014.185

[67] Musci TJ, Caughey AB. Cost-effectiveness analysis of prenatal population-based fragile X carrier screening. *Am. J. Obstet. Gynecol.* 2005;192:1905-12.

[68] Nolin SL, Brown WT, Glicksman A, Houck GE Jr, Gargano AD, Sullivan A, Biancalana V, Bröndum-Nielsen K, Hjalgrim H, Holinski-Feder E, Kooy F, Longshore J, Macpherson J, Mandel JL, Matthijs G, Rousseau F, Steinbach P, Väisänen ML, von Koskull H, Sherman SL. Expansion of the fragile X CGG repeat in females with premutation or intermediate alleles. *Am. J. Hum. Genet.* 2003;72:454-64.

[69] Rifé M, Badenas C, Quintó L, Puigoriol E, Tazón B, Rodriguez-Revenga L, Jiménez L, Sánchez A, Milà M. Analysis of CGG variation through 642 meioses in Fragile X families. *Mol. Hum. Reprod.* 2004;10:773-6.

[70] Lévesque S, Dombrowski C, Morel ML, Rehel R, Côté JS, Bussières J, Morgan K, Rousseau F. Screening and instability of FMR1 alleles in a prospective sample of 24,449 mother-newborn pairs from the general population. *Clin. Genet.* 2009;76:511-23.

In: Allelic Forms of the FMR1 Gene
Editor: Montserrat Milà
ISBN: 978-1-63321-914-4

Chapter 7

Treatment of Fragile X Spectrum: FXS, FXTAS and FXPOI

***F. J. Ramos*[1] *and M. P. Ribate*[2]**
[1]Unidad de Genética-Pediatría, GCV-CIBERER,
Hospital Clínico Universitario "Lozano Blesa",
Facultad de Medicina, Universidad de Zaragoza, Spain
[2]Facultad de Ciencias de la Salud, Universidad San Jorge,
Zaragoza, Spain

Abstract

Fragile X Spectrum includes three different clinical conditions Fragile X Syndrome (FXS), Fragile X-Associated Tremor and Ataxia (FXTAS), and Fragile X-Associated Premature Ovarian Insufficiency (FXPOI). Treatment of FXS is mainly symptomatic and it is addressed to improve or make disappear some of the more dyscapacitating symptoms like hyperactivity, deficit attention disorder and behavioral problems of language anomalies. Clinical trials are in course focusing on new discovered therapeutical targets (i.e. mGluR). Treatment of FXTAS and FXPOI are also mainly symptomatic and should be individually prescribed and modified depending on the clinical evolution.

Introduction

Fragile X spectrum includes Fragile X Syndrome (FXS), Fragile X-Associated Tremor Ataxia Syndrome (FXTAS) and Fragile X-Associated Premature ovarian Insufficiency (FXPOI). All three conditions are associated with the CGG trinucleotide expansion in the FMR1 gene on the X chromosome: FXS with more than 200 CGGs (full mutation) and FXTAS and FXPOI with 50 to 200 CGGs (premutation). Patients with FXS and FXTAS are mostly males, although females can also be affected. All three conditions are treated symptomatically, combining pharmacological and non-pharmacological therapies directed at alleviate the main disabling symptoms present in each condition.

Fragile X Syndrome

Fragile X syndrome (FXS) is the most common inherited cause of intellectual disability that affects approximately 1 in 4,000 males and 1 in 8,000 females. Its genetic cause was identified in 1991 and consisted of the expansion of the number of CGG trinucleotide repeats at the FMR1 gene, located on the distal region of the long arm of chromosome X. The normal gene has less than 55 CGGs but when it has more than 55 CGGs it tends to expand when transmitted by the mother to the next generation. Individuals who have between 55- 200 CGGs (premutation) are considered carriers, and those with more than 200 CGGs (full mutation) and have a hypermethylated gene are affected. This methylation leads to the silencing of the FMR1 gene and the subsequent absence, partial or complete, of the FMRP (Fragile X Mental Retardation Protein), considered as the ultimate cause of FXS [1]. The vast majority of FXS males with the full mutation develop intellectual disabilities, which occurs in only 25% of females due to the presence of a normal X chromosome that produces FMRP. However, most females with FXS have learning disabilities and/or emotional and behavior problems [2].

The clinical presentation of FXS is highly variable. In addition to intellectual disability, affected patients have neurodevelopmental problems such as attention deficit hyperactivity disorder, disruptive behavior or autism spectrum disorders. The physical characteristics of FXS include dysmorphic facial features (long face, large and prominent ears and prominent chin), and macroorchidism after puberty [3].

The FMRP Protein and the Synapse

FMRP is expressed ubiquitously in many tissues, predominantly in brain neurons. There it binds to different nuclear mRNAs, including the FMR1 gene mRNA, repressing the translation of the target mRNA to postsynaptic dendritic spines, where its activity is important for synaptic plasticity during transport. Dendritic spines are specialized and rapidly changing protrusions in the neurons whose creation and elimination are essential for learning and memory processing [4]. Microscopic analysis of brain samples of patients with FXS and Fmr1 knockout mice revealed no gross morphological abnormalities; however, in certain areas of the brain there are spindly dendritic spines, which are interpreted as immaturity and affect adversely the synaptic plasticity [5]. The discovery of a morphological spine phenotype indicates a possible defect in synaptic plasticity in FXS that could result in the intellectual disability phenotype. Whether the abnormal spine morphology is a cause or a consequence of altered signal transmission is currently unknown [6].

The Glutamate Receptor (mGluR) Hypothesis

In the brain, there are two main types of neurotransmitter receptors in the synaptic membrane: metabotropic receptors and ionotropic receptors. Metabotropic receptors (mGluRs) have eight different subtypes (mGluR1-8) which in turn are divided into three groups (I, II and III) according to sequence similarities and pharmacological properties. Group I includes mGluR1 and mGluR5 receptors which activates the Gq protein coupled and phospholipase C. Group II receptors (mGluR2 and mGluR3) and Group III (mGluR4, mGluR6, and Glu7 Glu8) are coupled to the proteins Gi / Go by inhibiting adenylate cyclase. Ionotropic receptors are ligand-gated ion channels in which the incorporation of a specific ligand induces a conformational change leading to opening of the receiver [7]. The opened receiver allows ion flow across the cell membrane by changing the excitability of the neuron. The main dependent neuronal glutamate ionotropic receptors are: AMPA (isoxazolepropionic amino-3-hydroxy-5-methyl-4-acid), NMDA (N-methyl-D-aspartic acid), and kainate receptors, whose activation produces a fast excitatory neurotransmission [8].

The mechanisms of long-term potentiation (LTP) and long-term depression (LTD) are the most important regulatory mechanisms of neuronal synaptic plasticity, defined as long-lasting changes in synaptic strength accompanied by abnormalities in the size and morphology of dendritic spines. The mechanism of strengthening of LTP produces a connection between presynaptic and postsynaptic neurons. The LTD mechanism is opposite to that of LTP and results in the weakening of synapses, mainly due to a reduction of glutamate-gated ionotropic AMPA-alpha in the postsynaptic membrane [9].

The 'theory of glutamate receptor (mGluR)', first published in 2004, tried to explain many aspects of clinical symptoms present in patients with FXS and in the Fmr1-KO mouse, including a higher density and immaturity of dendritic spines compared to normal individuals/mice. This theory stated that internalization of the AMPA receptor is excessive in FXS and likely caused by stimulation of group I mGluRs (mGluR1 and mGluR5). Such stimulation would induce local mRNA translation, resulting in the synthesis of new protein that triggers the internalization of the AMPA receptor, essential for the long-term plasticity of dendritic spines. FMR1 mRNA is also present in the postsynaptic compartment and FMRP is synthesized locally after mGluR activation [10].

On the other hand, FMRP seems to negatively regulate the translation of proteins that are important for the internalization of the AMPA receptor. The exact mechanism by which the FMRP local translation represses mRNA remains unknown. Moreover, phosphorylation and dephosphorylation of FMRP seems play a crucial role in the signaling cascade induced by stimulation of mGluR5 receptors, which stimulate local protein synthesis at the synapses [11].

Since the formulation of the mGluR theory, clinical researchers in FXS have sought the support of the pharmaceutical industry to collaborate in identifying therapeutic strategies aimed at mGluR5 in order to at least partially reverse some of the symptoms of FXS.

The MPEP (2-methyl-6-(phenylethynyl)-pyridine), a negative modulator of mGluR5, can, *in vitro*, counter the excess of activity of the receptor and rescue the loss of AMPA receptors after the loss of FMRP. In FXS animal models, researches have got the pharmacological rescue of audiogenic seizures, behavioral phenotypes and alterations in dendritic spines by using several negative modulators of mGluR5. MGluR theory has focused research into the basic mechanisms underlying FXS, prompting the search for new therapeutic strategies [12].

The GABA Hypothesis

Besides the mGluR′s hypothesis, researches have suggested that the signaling receptor gamma-aminobutyric acid (GABA) is altered in patients with FXS. GABA is the major inhibitory neurotransmitter in the central nervous system (CNS) and, as such, plays a key role in the modulation of neuronal activity in the brain. GABA mediates its action through two distinct systems, the ionotropic GABA-A and the metabotropic GABA-B receptors [13]. Many patients with FXS have epilepsy and sleep disorders, conditions associated with the signaling pathway of the GABA receptor. Interestingly, the mRNA encoding GABA-A receptor subunits are targets of FMRP. The Fmr1-KO mice have decreased levels of mRNA and protein in various subunits of GABA-A receptor compared to normal animals [14]. These studies support a model in which the FMRP mRNA would regulate the stability of the GABA-A receptor subunits, preventing its degradation by joining these mRNAs *in vivo*. In addition to GABA-A receptors, GABA-B receptors could also be related to FXS, therefore turning into another potential therapeutic target. The GABA-B receptor agonists inhibit the presynaptic release of glutamate and the postsynaptic signaling cascade downstream of mGluR5 [15].

All those studies have demonstrated a dysfunction in the mRNA and protein expression of several subunits of GABA receptors, mechanisms that would be likely involved in the occurrence of the FXS phenotype.

Therapeutic Approaches in FXS

Currently, the treatment of patients with FXS is mainly symptomatic. The two most commonly used medications are stimulants used to improve attention and hyperactivity, and selective inhibitors of serotonin reuptake to reduce the risk of aggression associated with anxiety.

FXS patients are not only treated with pharmacological agents, but also benefit from behavioral therapies for improving emotional and language problems. As already demonstrated in the mouse FXS behavior improves with an enriched environment and therefore this therapy might also be beneficial to humans [16]. Current therapeutic strategies, both pharmacological and non-pharmacological, are aimed at improving the symptoms but do not improve cognitive function. In recent years, new strategies have been developed for therapeutic interventions in FXS based on the theories of mGluR and GABA receptors. Several clinical trials using new designed drugs were initiated to correct the abnormal activity of the mGluR and GABA pathways in FXS [6]. Unfortunately, some of them have been recently discontinued due to the lack of demonstrable improvement.

MGluR5 Inhibitors

The *fenobam* is a potent and selective antagonist of mGluR5 and was the first negative mGluR5 modulator tested in patients with FXS. To test whether it had significant effect on the FXS phenotype, *fenobam* was administered as a single oral dose to twelve affected patients (six males and six females) that the pre-pulse flicker noise (PPI -Pre-Pulse inhibition was measured inhibition of Acoustic Startle-) before and after drug administration. Generally, patients with FXS show decreased PPI compared to healthy control individuals. Six of the twelve patients with FXS showed improvement after treatment with ~~PPI~~ *fenobam*, with no significant adverse effects observed [17].

Although results were promising, it was difficult to draw definitive conclusions because of the lack of a controlled study with placebo, the fact that the patients received only a single dose of *fenobam*, and the low number of participants enrolled. Other inhibitors are the AFQ056 mGluR5, from Novartis; the STX107, developed by Merck and whose license was issued to Seaside Therapeutics, and the RO4917523, from Hoffman-La Roche [6]. Despite of the results of a preliminary clinical trial with AFQ056 (*Mavoglurant*) in which certain cognitive aspects improved in FXS patients with complete methylation of the FMR1 gene [18], the trial was discontinued by Novartis last April after results of phase IIb/III failed to show demonstrable improvement. Last month, Roche discontinued its trial as well because of negative phase II clinical results.

GABA-A Agonists

GABA-A receptor agonists are currently used as anticonvulsants, antidepressants or anxiolytics. *Benzodiazepines,* which enhance the function of the GABA receptor, are the best-known drugs of this group. Although their anxiolytic effects are useful in patients with FXS have unwanted side effects such as sedation or ataxia, and treatment discontinuation may cause withdrawal symptoms [6]. Besides the selective agonists of GABA-A receptor, neuroactive steroids that allosterically modulate them may be also effective, for example *ganaxolone,* which has a favorable safety profile and may be useful in patients with FXS [19].

The use of inhibitors of GABA-A receptors in patients with FXS is likely to be effective in reducing symptoms such as seizures or sleep disorders.

GABA-B Agonists

Arbaclofen (STX209), a drug proven effective and safe for gastroesophageal reflux disease (GERD), has been used to treat autism spectrum disorders (ASD) in several studies. In an animal model of FXS, *arbaclofen* demonstrated a reversal of behavioral, neurological, and neuropathological features associated with the disease. Results from one double-blind, placebo-controlled study that treated children and adults with FXS have been promising, with signs of improvement in social function, especially in the most severely socially impaired individuals. Two studies, one open-label and one double-blind, placebo-controlled, were also conducted in children, adolescents, and young adults with ASD showing improvements in socialization [20, 21].

Approximately 13-18% of patients with FXS have seizures. The GABA-B and mGluR5 receptors are thought to be involved in the occurrence of audiogenic seizures in Fmr1-KO mice. Treatment of these mice with racemic baclofen reduced the incidence of seizures.

Another positive effect of GABA-B receptor agonists could be the reduction of anxiety in patients with FXS. Treatment of Fmr1-KO mice with a GABA-B receptors agonist can inhibit audiogenic seizures, supporting the idea that GABA-B receptors are involved in the etiology of FXS [6].

AMPA Agonists

The increased internalization of AMPA receptors in neurons of Fmr1-KO mice plays a major role in altering the transmission of signals. CX516 is an ampakine which acts as a positive allosteric AMPA receptor modulator. This compound binds to the AMPA receptor-channel complex and induces a slower deactivation of the receptor, which results in a longer opening time, in a slower disappearance of the excitatory postsynaptic potential, and ultimately to an increased hippocampal LTP, compared to baseline state before administering the drug. Thus, CX516 enhances AMPA receptors after glutamate-mediated synaptic. CX516 has been tested in phase II, randomized, double-blind, placebo-controlled safety for four weeks in adult patients with FXS. Unfortunately no significant or cognitive measures and improvement in behavior were observed. This could be due to the use of low doses, to the CX516 short half-life in humans or, finally, to insufficient time of treatment. In the study, there were only minimal side effects and no serious adverse effects were observed [22].

Theoretically, treatments targeting AMPA receptors could improve behavior in FXS, although the beneficial effects of ampakines have yet to be convincingly demonstrated.

NMDA Receptor Antagonists

Memantine is a non-competitive antagonist of NMDA receptors that can slow the progression of Alzheimer's disease and has been tested to treat Pervasive Developmental Disorders (PDD). In the presence of low levels of synaptic glutamate, binding of *memantine* blocks NMDA receptors, which are unblocked when glutamate levels increase. Besides excessive signaling through mGluR5 is linked to an increased AMPA receptor internalization and to the dysregulation of NMDA receptor activity [23]. Therefore, *memantine* may have positive effects on behavioral problems of patients with FXS.

A clinical trial of *memantine* in six patients with FXS who had a comorbid diagnosis of PDD has been reported. Therapeutic effects were determined by the clinical evaluation of the "Clinical Global Impressions-Improvement" (CGI-I) scale during the time of treatment. The results did not show any significant improvement, although in four of the six patients certain signs of improvement were observed. Furthermore, the study was not a randomized placebo-controlled trial and, therefore, it was difficult to draw valid conclusions [24].

The NMDA receptor may be a target for pharmacological treatment of SXF because some brain regions of the Fmr1-KO mice showed impaired NMDA-dependent LTP. However, there is no convincing evidence that NMDA receptor antagonists have demonstrable beneficial effects on behavior [6].

Additional Treatments

Lithium has been used for many years as a mood stabilizer. Most studies that have linked *lithium* with FXS have focused on the path of GSK3. *Lithium* inhibits the activity of GSK-3b, which in turn inhibits the phosphorylation of microtubule-associated protein 1B (MAP1B). The MAP1B is one of the major targets of the mRNA to which it binds and translationally regulates FMRP. In 2008, a clinical trial with open-label *lithium* was published in FXS patients; although the test was not randomized and placebo controlled, *lithium* appears that produced beneficial effects as decreased responses of aggression, abnormal vocalization improvement of self-harm and anxiety [25]. In conclusion, *lithium* appears to have beneficial effects on behavior and some cognitive functions but further research, such as long-term placebo-controlled studies are needed to check the effects of *lithium* in patients with FXS.

Minocycline is a tetracycline analog which can inhibit matrix metalloproteinase-9 (MMP-9) and reduce inflammation in the CNS. MMP-9 is an extracellular endopeptidase which breaks down the extracellular matrix proteins acting on synaptogenesis and morphology of dendritic spines. *Minocycline* is thought to act on the mGluR pathway and it has been tested in clinical trials to treat various neurological disorders (stroke, multiple sclerosis and autism). It has also been shown to have beneficial effects on the maturation of dendritic spines in cultured hippocampal neurons of Fmr1-KO mice [26]. A clinical trial of open-label studied the effects of *minocycline* in 50 patients with FXS was recently completed. The results were an improvement in language and behavior [27].

Acamprosate (calcium acetyl-homotaurine) is a commercially available drug used for the maintenance of abstinence from alcohol. This compound appears to have several mechanism of action: antagonist of mGluR5, weak antagonist of NMDA receptors and agonist of GABA-A receptors. Although the mechanism of action of *acamprosate* is not fully understood, a small clinical trial was conducted in three patients with FXS to evaluate the response to treatment with *acamprosate* using the CGI-I scale. After a minimum of 16 weeks of treatment, all three patients improved. Surprisingly, they also showed improvement in language skills. Two subjects experienced nausea. Although these results appear promising, this trial was not placebo-controlled and the number of participants was too small to draw definitive conclusions [28].

Aripiprazole, an atypical antipsychotic that was approved in the USA for the treatment of children and adolescents with autism, has also been tested in patients with FXS. The effect of *aripiprazole* in patients with FXS was assessed by the CGI-I and ABC-I scales. All subjects who completed the study showed significant improvement in irritable behavior [29]. However, the study was not placebo-controlled and therefore the findings were difficult to interpret.

Oxidative stress has been implicated in some psychiatric disorders, including autism. Studies in animal models suggested that in FXS there is an increased sensitivity to oxidative stress, which may alter neuronal and glial function. Indirect evidence was obtained by preclinical treatments of FXS performed using antioxidants. Indeed, chronic alpha-tocopherol treatment normalized free radical production and oxidative stress in the brain of FXS mice, in which it improved many of the behavioral and learning deficits of these mice, such as exploratory behaviors, habituation abnormalities, anxiety responses and contextual fear conditioning [30]. Recently, a phase II randomized, placebo-controlled trial was protocol was designed for children and adolescents with FXS [31].

Chronic melatonin treatment of FXS mice normalized the glutathione levels and prevented lipid peroxidation in the brain and testes of the mice. Moreover, melatonin improved some abnormal behaviors observed in FXS mice such as context-dependent exploratory and anxiety behaviors and learning abnormalities [32]. In a 4-week double-blind, placebo-controlled study with melatonin in 12 children with FXS, ASD or both, results showed a favorable effect (longer) in night sleep duration [33].

Final Remarks

FXS patients are treated with medications to alleviate the symptoms of the disease and/or the behavior problems such as hyperactivity or anxiety. Attempts to discover new drugs for the treatment of patients with FXS are becoming more promising with time due to improved understanding of the molecular mechanisms involved in the pathogenesis of the syndrome. Preliminary trials of new drugs in KO-mice models will be essential before they can be used in clinical trials in humans, although this does not guarantee their success.

To date, several clinical trials have been carried out with promising preliminary results. Nevertheless, others had to be discontinued due to lack of positive results. Hopefully, more trials are currently being conducted or designed with improvements such as randomized double-blind methodology, sufficient number of patients and the use of outcome objective measurements to determine the therapeutic efficacy of the drug. For this, most clinical trials used one or more classic psychological questionnaires, however, the measurements of improvement were based on the likely subjective assessments made by relatives, caregivers and teachers. To ensure objectivity, it is important to develop reliable and objective outcome measurements (i.e.: questionnaires, scales or tests) to determine the true efficacy of the new drug [34]. Fortunately, and despite of some recent disappointing results, a number of clinical-therapeutic trials for FXS and/or ASD are currently in course or being designed.

It is remarkable that in less than 25 years since the discovery of the genetic defect that causes FXS, targeted therapeutic strategies have been already developed, with more or less success, to treat some of the most disabling symptoms of affected individuals. Although curative treatment seems to be far yet, it is likely that in the near future more effective symptomatic treatments will be available for FXS, especially for individuals with ASD. If expectations are met, they will significantly improve their quality of life ~~of them~~ and of their families.

Fragile X-Associated Tremor-Ataxia Syndrome (FXTAS)

Fragile X-Associated Tremor-Ataxia Syndrome (FXTAS) is a progressive neurological disease that affects mainly males over 50 who carry a premutation allele (55-200 CGG repeats) in the fragile X gene FMR1. Affected individuals usually have intention tremor, ataxia, signs of parkinsonism, cognitive progressive decline and peripheral neuropathy. Besides, ancillary findings include autonomic dysfunction and psychiatric symptoms such as anxiety, depression and disinhibition [35].

As in FXS, there is no curative treatment for FXTAS, although there are a number of symptomatic therapies that can improve the clinical manifestations in affected individuals. Moreover, there is enough evidence regarding the efficacy of various medications for treatment of other diseases (i.e. Alzheimer diseases) that have important clinical overlap with FXTAS.

The most common therapeutic interventions for FXTAS are listed in Table 1, which includes symptom-based treatments [36]. The drugs are used, basically, to alleviate symptoms related to tremor, equilibrium coordination, parkinsonism, sleep problems, anxiety, mood alterations, memory problems and pain.

Fragile X-Associated Primary Ovary Insufficiency (Fxpoi)

Primary ovarian insufficiency or premature ovarian failure (POI/POF) is one of the causes of female infertility. POI consists in cessation of menstrual periods, increased levels of FSH and diminished levels of estrogens, all before the age of 40. POI occurs in about 1% of women between 30 to 40 years of age. Fertility of women with POI is severely diminished, but unlike menopause, POI may be accompanied with spontaneous ovarian activity and natural pregnancies [37].

The major causes of POI include autoimmunity, genetic and environmental factors. Among the most common genetic conditions that produce POI is Fragile X premutation, which is present in all female carriers.

Treatment of POI includes hormone replacement therapy to reduce complications due to impaired endocrine function of ovaries, and fertility preservation therapies that include ovarian cortex, oocyte and embryo cryopreservation, oocyte or embryo donation and adoption in women without any ovarian function [38].

Recent research findings in animals and humans showed that neonatal and adult ovaries have some oogonial stem cells (OSCs) that can stably proliferate for months and produce mature oocytes *in vitro*. Studies on the isolation of OSCs form ovaries of aged animals and production of mature normal oocytes in ovaries of young adult animals lead to the recognition of the ~~of~~ importance of OSC niche and intraovarian environment on their differentiation to mature, normal oocytes. Therefore, cases of POI that result from defects in ovarian niche and its incapacity to support differentiation and growth of oocytes and also ovarian aging may be reversible in future [39].

These results have offered the opportunity for the application of OSCs as a target for POI therapy, restoration of ovarian function and, subsequently, restoration of normal fertility. However, clinical utility of these cells for treatment requires more evidence to confirm their safety, especially the effects from epigenetic changes during *in vitro* culture, and manipulation of produced oocytes and also resultant offspring.

Table 1. Symptom-based treatments for patients with FXTAS

Symptoms	Treatment/Therapy
Tremor	Primidone, beta-blockers, benzodiazepines
Ataxia	Amantadine and physical therapy
Parkinsonism	Carbidopa/levodopa, pramipexole and eldepryl
Cognitive deficits and dementia	Donepezil, rivastigmine, galantamine, memantine
Psychiatric problems	Sertraline, citalopram, escitalopram, duloxetine, mirtazapine, venlafaxine and aripiprazole
Autonomic dysfunction	Bladder incontinency: Tricyclic antidepressants, muscarinic receptor antagonists, cytoscopy with injection of Botox Swallowing difficulties: pyridostigmine bromide
Pain	Antidepressants, antiepileptics, topical analgesics, gabapentin and/or pregabalin

*Modified from Capelli et al., (2010). [36]

Conclusion

There is no curative treatment for Fragile X Syndrome. Symptomatic paliative therapies include behavioral and cognitive interventions, speech therapy, occupational and physical therapy, as well as pharmacological treatments. All of them intended to improve intellectual abilities, familial interactions and social integration. Pharmacological treatments should be individualized depending on the age and the intellectual-cognitive level of the affected individual. Treatment for Fragile X-Associated Tremor and Ataxia is also symptomatic, and include physical therapy, psychological and/or psychiatric interventions, and pharmacological treatments. Fragile X Carrier females with Premature Ovarian Insufficiency should be treated by the appropriate specialists. Hormone replacement and fertility preservation therapies are the two main issues that should be addressed.

References

[1] Oostra, B. A. and Willemsen, R. FMR1: A gene with three faces. *Biochim Biophys Acta* 2009, 1790, 467-477.

[2] Garber, K. B., Visootsak, J., Warren, S. T. Fragile X Syndrome. *Eur J Hum Genet* 2008, 16, 666-672.

[3] Brown W. T. *Clinical aspects of the Fragile X Syndrome.* In: Denman R. B. Ed. Modeling Fragile X Syndrome: Results and Problems in Cell Differentiation nº 54. Berlin: Springer-Verlag; 2012; pp. 273-279.

[4] Kim, M., Bellini, M., Ceman, S. Fragile X mental retardation protein FMRP binds mRNAs in the nucleus. *Mol Cell Biol* 2009, 29, 214-228.

[5] Irwin, S. A., Patel, B., Idupulapati, M., Harris, J. B., Crisostomo, R. A., Larsen, B. P., Kooy, F., Willems, P. J., Cras, P., Kozlowski, P. B., Swain, R. A., Weiler, I. J., Greenough, W. T. Abnormal dendritic spine characteristics in the temporal and visual cortices of patients with fragile-X syndrome: A quantitative examination. *Am. J Med Genet* 2001, 98, 161-167.

[6] Levenga, J., de Vrij, F. M. S., Oostra, B. A., Willemsen, R. Potential therapeutic interventions for Fragile X síndrome. *Trends Mol Med* 2010, 16, 516-527.

[7] Gross, C., Berry-Kravis, E. M., Bassell, G. J. Therapeutic Strategies in Fragile X Syndrome: Dysregulated mGluR Signaling and Beyond. *Neuropsychopharmacology* 2011, 37, 178-195.

[8] Ferraguti, F., Crepaldi, L., Nicoletti, F. Metabotropic glutamate 1 receptor: Current concepts and perspectives. *Pharmacol Rev* 2008, 60, 536-581.

[9] Collingridge, G. L., Peineau, S., Howland, J. G., Wang, Y. T. Long-term depression n the CNS. *Nat Rev Neurosci* 2010, 11, 459-473.

[10] Bear, M. F., Huber, K. M., Warren S. T. The mGluR theory of fragile X mental retardation. *Trends Neurosci* 2004, 27, 370-377.

[11] Ceman, S., O´Donnell, W. T., Reed, M., Patton, S., Pohl, J., Warren, S. T. Phosphorylation influences the translation state of FMRP-associated polyribosomes. *Hum Mol Genet* 2003, 12, 32295-32305.

[12] Nakamoto, M., Nalavadi, V., Epstein, M. P., Narayanan, U., Bassell, G. J., Warren, S. T. Fragile X mental retardation protein deficiency leads to excessive mGluR5-dependent internalization of AMPA receptors. *Proc Natl Acad Sci USA* 2007, 104, 15537-15542.

[13] D'Hulst, C. and Kooy, R. F. The GABA(A) receptor: A novel target for treatment of fragile X? *Trends Neurosci* 2007, 30, 425-431.

[14] Selby, L., Zhank, C., Sun, Q. Q. Major defects in neocortical GABAergic inhibitory circuits in mice lacking the fragile X mental retardation protein. *Neurosci Lett* 2007, 412, 227-232.

[15] Tu, H., Rondard, P., Xu, C., Bertaso, F., Cao, F., Zhang, X., Pin, J. P., Liu, J. Dominant role of GABAB2 and Gbetagamma for GABAB receptor-mediated-ERK1/2/CREB pathway in cerebellar neurons. *Cell Signal* 2007, 19, 1996-2002.

[16] Meredith, R. M., Holmgren, C. D., Weidum, M., Burnashev, N., Mansvelder, H. D. Increased threshold for spike-timing-dependent plasticity is caused by unreliable calcium signaling in mice lacking fragile X gene FMR1. *Neuron* 2007, 54, 627-638.

[17] Hessl, D., Berry-Kravis, E., Cordeiro, L., Yuhas, J., Ornitz, E. M., Campbell, A., Chruscinski, E., Hervey, C., Long, J. M., Hagerman, R. J. Prepulse inhibition in fragile X syndrome: feasibility, reliability, and implications for treatment. *Am J Med Genet B Neuropsychiatr Genet* 2009, 150B, 545-553.

[18] Jacquemont, S., Curie, A., des Portes, V., Torrioli, M. G., Berry-Kravis, E., Hagerman, R. J., Ramos, F. J., Cornish, K., He, Y., Paulding, C., Neri, G., Chen, F., Hadjikhani, N., Martinet, D., Meyer, J., Beckmann, J. S., Delange, K., Brun, A., Bussy, G., Gasparini, F., Hilse, T., Floesser, A., Branson, J., Bilbe, G., Johns, D., Gomez-Mancilla, B. Epigenetic modification of the FMR1 gene in Fragile X Syndrome is associated with differential response to the mGluR5 antagonist AFQ056. *Sci Transl Med* 2011; 3:64ra1.

[19] Cornish, K., Turk, J., Hagerman, R. The fragile X continuum: New advances and perspectives. *J Intellect Disabil Res* 2008, 52, 469-482.

[20] Cryan, J. F. and Kaupmann, K. Don't worry 'B' happy! A role for GABA(B) receptors in anxiety and depression. *Trends Pharmacol Sci* 2005, 26, 36-43.

[21] Erickson, C. A., Veenstra-Vanderweele, J. M., Melmed, R. D., McCracken, J. T., Ginsberg, L. D., Sikich, L., Scahill, L., Cherubini, M., Zarevics, P., Walton-Bowen, K., Carpenter, R. L., Bear, M. F., Wang, P. P., King, B. H. STX209 (*arbaclofen*) for autism spectrum disorders: An 8-week open-label study. *J Autism Dev Disord* 2014, 44, 958-964.

[22] Berry-Kravis, E., Krause, S. E., Block, S. S., Guter, S., Wuu, J., Leurgans, S., Decle, P., Potanos, K., Cook, E., Salt, J., Maino, D., Weinberg, D., Lara, R., Jardini, T., Cogswell, J., Johnson, S. A.; Hagerman, R. Effect of CX516, an AMPA-modulating compound, on

cognition and behavior in fragile X syndrome: a controlled trial. *J Child Adolesc Psychopharmacol* 2006, 16, 525-540.

[23] Shang, Y., Wang, H., Mercaldo, V., Li, X., Chen, T., Zhuo, M. Fragile X mental retardation protein is required for chemically-induced long-term potentiation of the hippocampus in adult mice. *J Neurochem* 2009, 111, 635-646.

[24] Erickson, C. A., Mullett, J. E., McDougle, C. J. Open-label *memantine* in fragile X syndrome. *J Autism Dev Disord* 2009, 39, 1629-1635.

[25] Berry-Kravis, E., Sumis, A., Hervey, C., Nelson, M., Porges, S. W., Weng, N., Weiler, I. J., Greenough, W. T. Open-label treatment trial of *lithium* to target the underlying defect in fragile X syndrome. *J Dev Behav Pediatr* 2008, 29, 293-302.

[26] Bilousova, T. V., Dansie, L., Ngo, M., Aye, J., Charles, J. R., Ethell, D. W., Ethell, I. M. *Minocycline* promotes dendritic spine maturation and improves behavioural performance in the fragile X mouse model. *J Med Genet* 2009, 46, 94-102.

[27] Utari, A., Chonchaiya, W., Rivera, S. M., Schneider, A., Hagerman, R. J., Faradz, S. M., Ethell, I. M., Nguyen, D. V. Side effects of *minocycline* treatment in patients with fragile X syndrome and exploration of outcome measures. *Am. J Intellect Dev Disabil* 2010, 115, 433-443.

[28] Erickson, C. A., Mullett, J. E., McDougle, D. J. Brief report: *Acamprosate* in fragile X syndrome. *J Autism Dev Disord* 2010, 40, 1412-1416.

[29] Erickson, C. A., Stigler, K. A., Wink, L. K., Mullett, J. E., Kohn, A., Posey, D. J., McDougle, C. J. A prospective open-label study of *aripiprazole* in fragile X syndrome. *Psychopharmacology* (Berl). 2011, 216, 85-90.

[30] De Diego-Otero, Y., Romero-Zerbo, Y., El Bekay, R., Decara, J., Sánchez, L., Rodriguez-de-Fonseca, F., del Arco-Herrera, I. Alpha-tocopherol protect against oxidative stress in the fragile X knockout mouse: An experimental therapeutic approach for the Fmr1 deficiency. *Neuropsycopharmacology* 2009, 34, 1011-1026.

[31] De Diego-Otero, Y., Calvo-Medina, R., Quintero-Navarro, C., Sánchez-Salido, L., García-Guirado, F., Del Arco-Herrera, I., Fernández-Carvajal, I., Ferrando-Lucas, T., Caballero-Andaluz, R., Pérez-Costillas, L. A combination of ascorbic acid and α-tocopherol to test the effectiveness and safety in the fragile X syndrome: study protocol for a phase II, randomized, placebo-controlled trial. *Trials* 2014, 15, 345-360.

[32] Romero-Zerbo, Y., Decara, J., el Bekay, R., Sanchez-Salido, L., Del Arco-Herrera, I., de Fonseca, F. R., de Diego-Otero, Y. Protective effects of melatonin against oxidative stress in Fmr1 knockout mice: a therapeutic research model for the fragile X syndrome. *J Pineal Res* 2009, 46, 224-234.

[33] Wirojanan, J., Jacquemont, S., Diaz, R., Bacalman, S., Anders, T. F., Hagerman, R. J., Goodlin-Jones, B. L. The efficacy of melatonin for sleep problems in children with autism, fragile X syndrome, or autism and fragile X syndrome. *J Clin Sleep Med* 2009, 5, 145-150.

[34] Berry-Kravis, E., Hessl, D., Abbeduto, L., Reiss, A. L., Beckel-Mitchener, A., Urv, T. K. and Outcome Measurements Working Groups. Outcome measures for clinical trials in Fragile X Syndrome. *J Dev Behav Pediatr* 2013, 34, 508-522.

[35] Hagerman, R. J., Hall, D. A., Coffey, S., Leehey, M., Burgeois, J., Gould, J., Zhank, L., Seritan, A., Berry-Kravis, E., Olichney, J., Miller, J. W., Fong, A. L., Carpenter, R., Bodine, C., Gane, L. W., Rainin, E., Hagerman, H., Hagerman, P. J. Treatment of fragile X-associated tremor ataxia syndrome (FXTAS) and related neurological problems. *Clin Intervention Aging* 2008, 3, 251-262.

[36] Capelli, L. P., Rodrigues-Gonçalves, M. R., Leite, C. C., Barbosa, E. R., Nitrini, R., Vianna-Morgante, A. M. The fragile X-associated tremor and ataxia syndrome (FXTAS). *Arq Neuropsiquiatr*, 2010, 68, 791-798.

[37] Sadeghi, M. R. New hopes for the treatment of Primary Ovarian Insufficiency/Premature Ovarian Failure. *J Reprod Infertil* 2013, 14, 1-2.

[38] Blumenfeld, Z. Fertility treatment in women with premature ovarian failure. *Expert Rev Obstet Gynecol* 2011, 6, 321-330.

[39] White, Y. A., Woods, D. C., Takai, Y., Ishihara, O., Seki, H., Tilly, J. L. Oocyte formation by mitotically active germ cells purified from ovaries of reproductive-age women. *Nat Med* 2010, 18, 413-421.

Index

A

B

C

D

E

F

G

H

I

J

K

L

M

Q

R

S